ONE-OF-A-KIND RESEARCH AIRCRAFT

A HISTORY OF IN-FLIGHT SIMULATORS, TESTBEDS & PROTOTYPES

Steve Markman & Bill Holder

Schiffer Military/Aviation History
Atglen, PA

Acknowledgements

As could be expected with any project, we had help, and lots of it. People we contacted for information on various aircraft were eager to assist by providing information and pictures, giving us leads to other people, and sometimes even reviewing our rough drafts. Our sincerest thanks and appreciation to everyone who was so eager to help, and also to those who were not so eager, but pitched in nonetheless.

Jim Acree, University of Tennessee Space Institute
Joyce Baker, History Office, Edwards Air Force Base
John F. Ball, Calspan Corp
A. Finley Barfield, Wright Laboratory, U.S. Air Force
David Beasley, Lockheed Corp
Dr Sergey Boris, Gromov Flight Research Institute, Zhukovsky, Russia
Sqn Ldr Peter Chandler, Empire Test Pilots' School, Royal Air Force
Bruce J. Csaszar, Lockheed Corp
Tom Drennen, Sikorsky Aircraft Corp
Dietrich Hanke, Flight Mechanics Institute, Braunschweig, Germany
Dr Gavin Jenney, Dynamic Controls, Inc
Fred Johnson, History Office, Edwards Air Force Base
James Kocher, Wright Laboratory, U.S. Air Force
Robert E. Lemble, Wright Laboratory, U.S. Air Force
Dave Menard, United States Air Force Museum
Don Nolan, NASA, Dryden Flight Research Center
John Perdzock, Wright Laboratory, U.S. Air Force
Duane P. Rubertus, Wright Laboratory, U.S. Air Force
Yasuo Sakurai, Japan Defense Agency, Tokyo, Japan
Mary F. Shafer, NASA, Dryden Flight Research Center
Sqn Ldr David Southwood, Empire Test Pilots' School, Royal Air Force
Frank Swortzel, Wright Laboratory, U.S. Air Force
Brian W. Van Vliet, Wright Laboratory, U.S. Air Force
Todd T. Wilkinson, General Electric Corp

Our special thanks to Morris A. Ostgaard for writing the Foreword to this book. For over thirty years, Morry was a fixture at the Wright Laboratory, and was the resident expert in flight control technology – always on the cutting edge of new ideas, and leading the way. Like many of the aircraft in this book, Morry is unknown outside the industry, but a legend within it.

And an extra special thanks to our loving wives, Ruthanne Holder and Helen Markman, who forgave us many times while we were typing feverishly at our computers or frantically reviewing stacks of information, rather than doing other things.

Book Design by Robert Biondi.

Copyright © 1995 by Steve Markman & Bill Holder.
Library of Congress Catalog Number: 95-67628

All rights reserved. No part of this work may be reproduced or used in any forms or by any means – graphic, electronic or mechanical, including photocopying or information storage and retrieval systems – without written permission from the copyright holder.

Printed in Hong Kong.
ISBN: 0-88740-797-8

We are interested in hearing from authors with book ideas on related topics.

Published by Schiffer Publishing Ltd.
77 Lower Valley Road
Atglen, PA 19310
Please write for a free catalog.
This book may be purchased from the publisher.
Please include $2.95 postage.
Try your bookstore first.

FOREWORD

Flight research is an integral and inseparable part of the aircraft design process. Aircraft of today probably would still resemble the original Wright Brother's aircraft were it not for the dedicated men and women who developed new concepts for aircraft, and then refined them, and then tested them in actual aircraft. Many types of aircraft have been developed to test new designs. Some were designed and built to demonstrate the feasibility of a specific new technology. These were the X series aircraft, such as the X-1, X-15, and X-29. Without these aircraft, many new ideas would have remained just that . . . ideas. These aircraft demonstrated quantum leaps in technology. Many books have been written to document their contributions.

There is a large number of lesser known and seldom recognized test aircraft that also made significant contributions to flight research. These are the testbeds and in-flight simulators that tested new materials, validated new aerodynamic shapes, tested new flight control concepts, determined and validated proper flying qualities, and developed new types of cockpit displays and controls. These aircraft made essential contributions to the design of new aircraft. Even before many of the X series aircraft flew, individual elements of these aircraft flew on other research aircraft.

The X series aircraft always captured our enthusiasm, as well as the headlines and book titles. Although known within the aircraft development and test communities, most other research aircraft are less well known by the general public. This book is dedicated to those aircraft. The aircraft in this book made contributions that were no less significant, and their places in history are equally deserved. At long last a book documents their contributions.

Unfortunately, this book is not complete. There are many more research aircraft that have been operated over the years. If you have information about any of them, or you had personal experiences with any of the aircraft already in this book, please feel free to contact either of the authors through the publisher. Certainly, with your help, more volumes of this book can be written.

Morris A. Ostgaard
Aerospace Consultant

CONTENTS

Acknowledgements 2
Foreword 3
Preface 5
Introduction 7

PART I: Inflight Simulator Aircraft 8
- VISTA/NF-16D 10
- Variable Stability B-26 14
- NC-131H Total In-Flight Simulator 16
- Gulfstream Shuttle Training Aircraft 20
- ASTRA Hawk 22
- University of Tennessee Navions 23
- P-2 Variable Stability Aircraft 24
- S-76 SHADOW 25
- NT-33A 28
- Tu-154M 31
- VFW-614 ATTAS 33
- Calspan Learjets 34
- Jetstar GPAS 36

PART II: Testbed Aircraft 40
- A-5A Vigilante SST 41
- A-6A CCW 42
- B-47 Fly-By-Wire 44
- A-7 DIGITAC 46
- B-52 CCV/LAMS 48
- Carrier Testbeds 51
- XC-8A ACLS 57
- Convair-990 LSRA 59
- C-130 RAMTIP 61
- Falcon ATLAS 62
- F-4 Fly-by-Wire 63
- F-5D Skylancer Testbed 66
- F-8 Supercritical Wing 67
- F-8 Digital Fly-By-Wire 68
- F-15 AECS 71
- F-15 ASAT 72
- F-15 IFFC/ABICS/ICAAS 74

- F-15 HIDEC 77
- F-15 STOL/MTD, ACTIVE 78
- F-15 Streak Eagle 81
- F-16 AFTI 82
- F-16 CCV, FLOTRAK 86
- F/A-18 EPAD 89
- F/A-18 HARV 90
- F/A-18 SRA 92
- JF-100 Variable Stability Testbed 94
- F-102 Low L/D 95
- F-104 Low L/D 96
- F-104 Aerospace Trainer 99
- F-100/106 Turbulance Testing 101
- F-111 AFTI/TACT Testbed 102
- Air Force Transport Testbeds 106
- Ice Testing Aircraft 113
- KC-135 Winglet 114
- NASA/Langley Commercial Testbeds ... 115
- L-100 High Technology Testbed 117
- PA-30 Twin Commanche Testbed 119
- Sabreliner Supercritical Wing 121
- SR-71 Testbed 122
- Boeing 737 TCV 123
- Boeing 720 CID 125
- X-21 LFC 128
- YF-23 Loads 130
- Miscellaneous Testbeds 132

PART III: Prototype Aircraft 136
- YA-7F (A-7 Plus) 137
- F-16XL 140
- F-16/79/101 142
- P-51 Mustang-Based Enforcer 144
- Gunships 146
- F-15E Strike Eagle Demonstrator 148
- F-18 149
- A-37 150

PREFACE

How did this book come to be written? It's a long story. Well actually, it's a short story.

Back in the early 1970s I was a young aerospace engineer just out of college. I had a new job, new home, new wife, and lots of dreams for the future. One of them was the plan some day to own my own World War II fighter. I dreamed big back in those days! I also read lots of aviation magazines back then (I had lots of time, too!), especially ones about the great fighters and bombers of World War II. A name I used to see on a lot of articles was Bill Holder. Bill could make those old aircraft come to life . . . almost like I could jump into the cockpits and put my hands on the controls. Somewhere I learned that Bill also lived in Dayton and also worked at Wright-Patterson Air Force Base. We may very well have passed each other many times over the years without ever knowing it.

Eventually my interests turned elsewhere as I realized I'd never own that Mustang or Thunderbolt. My work eventually led me into the fascinating business of flight testing . . . not as a pilot, but as the program manager for the Air Force's fleet of in-flight simulator aircraft. Projects often led me to aircraft flight test centers, aircraft factories, and modification centers throughout the country and overseas. I saw lots of aircraft modified for test programs and met lots of fascinating people throughout the flight test business. I wish I had a dollar for every meeting I attended with some famous test pilot or astronaut.

I often browsed through aviation books at the library and book stores. There were many books about the X-1, X-15, and X-29, but I seldom saw any mention of many of the aircraft I had seen. I often thought someone was missing a great opportunity to tell this fascinating story to the world. Although I enjoyed writing, I felt such an effort was beyond me.

Jump ahead a few years to about 1990. Late one Friday night I was on a flight home to Dayton from Washington. Trying to get comfortable, I happened to notice the name on the brief case of the man seated next to me . . . Bill Holder. Finally summoning the courage, I asked, "Excuse me, but are you Bill Holder the aviation writer?" To make a long story short, obviously he was. We chatted most of the way home, during which time I talked about my book idea. Bill became fascinated with the idea also. By the time the flight ended, we had roughed out an outline and agreed to pursue the idea.

The rest, so to say, is history.

Steve Markman

One of the biggest users of one-of-a-kind research aircraft is the NASA Langley Research Center. The organization's interesting fleet is displayed here. (NASA Photo)

INTRODUCTION

Test aircraft are nothing new – they have been around since the first aircraft flew. Indeed, the first airplane was a test airplane. No one could guarantee for sure that it would work – someone had to attempt to take it into the air to find out. The first powered aircraft of the Wright Brothers, and the gliders of Langley, Lilienthal, and others that preceded the Wrights were test aircraft, testing the ideas of their designers.

The modern production aircraft of today, the F-15s, 767s, etc, developed as a result of a sometimes slow, sometimes fast evolution from the first powered aircraft. An essential part of this evolution is the use of test aircraft to develop and verify new capabilities. Many excellent books have been written about test aircraft such as the X-1 and X-15 that have made great leaps in technology. These aircraft demonstrated quantum jumps in technology, proving that aircraft can go faster than the speed of sound, or take man to the edge of space. They showed that big steps can be taken and paved the way for later production aircraft that would incorporate the technology they validated. After demonstrating the "great leap" in technology for which they were designed, these aircraft usually were retired and their designers and pilots moved on to other challenges. These aircraft rightfully earned their places in honored displays in museums and in the pages of history books.

But there are other, lesser known, test aircraft that seldom demonstrate the great leap in technology. They do the day-to-day research, making small gains a little at a time, slowly nibbling away at the envelope, paving the way for new production aircraft or for other test aircraft that

The use of production aircraft for simulation and testbed missions has been carried out for many years. Here, a C-47 assigned to the Air Force Medical Research Laboratory was used in the 1950s for human engineering experiments. (USAF Photo)

will make the "great leap." These aircraft are used to test new materials, validate new aerodynamic shapes, test new flight control concepts, determine proper flying qualities, and develop new types of cockpit displays. These aircraft, doing the grunt work but seldom receiving the notoriety of their more famous hangar mates, are the aircraft this book will discuss.

The important difference with the aircraft in this book is that these were production aircraft that were modified to serve as test beds, technology demonstrators, or in-flight simulators. They are modified as needed to look at a specific technology improvement, tested, then modified again to perform other research. Many have been in use for decades. Although seldom written about except in technical journals, the research they perform constitutes much of the data base upon which new aircraft are designed. This book constitutes the first publication aimed at the general public to tell the story of many of these unique research aircraft.

PART I
IN-FLIGHT SIMULATORS

Have you ever wondered what it would be like to fly a new aircraft even before it is built? How will it fly? Will it roll just right or be too abrupt or too sluggish? Will it sink too fast as the pilot begins the landing flare? Will it become uncontrollable in turbulence?

To an extent, analysis techniques that aircraft designers use answer some of these questions. However, they leave out one very important piece of information ... how the pilot will interact with the aircraft. Considering the piloted airplane as an entire system, the pilot is the most unpredictable part. No two pilots are alike, and any one pilot's response can change depending on aspects such as how long his flight is, the dangers he will face, how much sleep he had the night before, and even whether or not his wife is mad at him.

Flight simulators help the aircraft designer to compensate for the different pilots' capabilities by putting the pilot into the design process. In a flight simulator, the pilot sits in a replication of the cockpit, sees the world through a computer generated visual system, and feels the motion of the aircraft through a sophisticated motion system to move the cockpit in response to his control inputs. The pilot experiences the flight characteristics and evaluates them as good, bad, or in between, just as in the real aircraft. Many new aircraft designs have been evaluated using flight simulators, and untold millions of dollars (and probably a few lives) have been saved by finding problems before the aircraft was ever built.

But despite all this sophisticated technology to make the pilot feel like he is in a real airplane, flight simulators still cannot do one thing ... convince the pilot that he *really* is in flight. Experience has proven that pilots fly a flight simulator differently because they know they cannot damage an aircraft or get hurt, even if they slam into the runway too hard or collide with another aircraft in flight. Because they know they cannot get hurt, pilots fly flight simulators more relaxed, take more chances, and don't initiate corrective actions until later, as compared to flying a real airplane. There have been many examples where the real aircraft experienced problems that were completely unpredicted after extensive flight simulation. The reason is that the pilots were too relaxed, and the problems went undetected. The pilot usually has to be very aggressive and abrupt to bring out the more subtle types of flying problems. A very relaxed pilot usually will not discover subtle problems.

So, enter the in-flight simulator. What is an in-flight simulator? It is very special type of research aircraft whose flight characteristics can be changed to match those of another aircraft. Just as with a flight simulator, the in-flight simulator can be made to duplicate the flight characteristics of aircraft that have not yet been built. This aircraft can be either a specific one being designed, or a theoretical "paper design" being evaluated for research purposes only. The in-flight simulator overcomes the three biggest limitations of ground-based flight simulators: first, the pilot sees the real world in three dimensions as he looks out the cockpit rather than looking at a flat computer generated view; second, the motions are real and sustained rather than the limited motion felt on the ground; and third, he now knows he is in a real aircraft, and any mistake can result in a broken aircraft, personal injury, or worse. Flying an in-flight simulator, the pilot is flying a real aircraft and acts accordingly.

In-flight simulators perform a variety of specialized missions which can be broken into four general groups:

- Pre-First Flight Evaluation of New Aircraft: Prior to first flight, new aircraft are simulated on an in-flight simulator to look for any previously undiscovered flying qualities problems. If there are none, then the designers' and test pilots' confidence in the new design increases. If there are problems, fixes can often be designed using the in-flight simulator and tested quickly.

PART I: IN-FLIGHT SIMULATORS

- Research and Development: Generic research in the areas of aircraft flight control, display development, human factors, and even medical research is performed using in-flight simulators. In addition, the development of new aircraft is supported by evaluating design concepts early in the design process, before designs are frozen, when changes can be made most easily and cheaply.

- Specialized Training: Space Shuttle astronauts and new test pilots train routinely using in-flight simulators. It is safer and cheaper to learn to fly the Space Shuttle using an in-flight simulator made to feel and fly like the Space Shuttle than for the astronaut to have to figure it out for real on his first Shuttle flight.

- Flying Laboratory: Because most in-flight simulators are highly instrumented, they make ideal platforms for testing new equipment and trying new ideas.

A detailed discussion of how in-flight simulators work is beyond the scope of this book. However, in simple terms, the pilot's control stick inputs go to a computer on which the flight characteristics of the simulated aircraft have been programmed. The computer figures out how the simulated aircraft would respond to the pilot's inputs and moves the in-flight simulator's control surfaces to produce the desired motions.

The in-flight simulators also must have a control stick or wheel and rudder pedals whose feel characteristics can be changed to duplicate the feel of the aircraft being evaluated. With these accomplished, the in-flight simulator responds to the pilot's commands the way the aircraft being simulated would respond.

But the changes do not stop there. It is also necessary to configure the cockpit to replicate, to the extent necessary, the cockpit environment and instrument displays. On some in-flight simulators, the instruments can be rearranged and/or electronic displays can be programmed to provide specific graphic features. Properly sized control sticks or control wheels can be installed and canopies can be masked to produce the same view out of the cockpit.

Most in-flight simulators have at least two pilots: an evaluation pilot who flies the aircraft being simulated, and a safety pilot who flies the in-flight simulator itself and can take control if the simulated aircraft proves too difficult for the evaluation pilot to fly. When the safety pilot takes control, he is flying the host, not the simulated aircraft. Because of this, new unknown designs, or designs with known problems can be simulated safety.

In-flight simulators have been in use since the early 1950s. Nearly all were developed from production aircraft that already had the performance and strength needed and could be modified to incorporate the new capabilities. Although not well known outside the aircraft design and flight test industries, they have been a valuable tool for designers and will continue to be used for many years to come.

ONE-OF-A-KIND RESEARCH AIRCRAFT

VISTA/NF-16D Variable Stability In-Flight Simulator

AIRCRAFT TYPE: F-16D
MISSION: In-Flight Simulation
OPERATING PERIOD: 1993 - Present
RESPONSIBLE AGENCY: Wright Laboratory, Wright-Patterson AFB, Ohio
CONTRACTOR: Lockheed Corporation, Calspan Corporation

THE STORY
The Variable Stability In-Flight Simulator Test Aircraft, or VISTA, was built as a new in-flight simulator that would have performance representative of the latest operational fighter aircraft. Built around a production F-16D airframe, the VISTA comprises production features from many different F-16 variants, plus an complex custom-designed variable stability system to allow the VISTA's flight characteristics to be changed.

The history of the VISTA goes back at least to the mid-1960s, when the need to replace the NT-33 in-flight simulator was identified. The NT-33A's performance was not representative of the latest fighter aircraft in use, nor of those that would be developed soon. Also, the future of the T-33 fleet was uncertain, and operating an aircraft without depot support would be very difficult and expensive (it is ironic that the NT-33A remains in service over twenty five years later). Several studies were performed in the early 1970s to develop an NT-33A replacement, one even culminating in the conceptual design of an F-4 in-flight simulator. Unfortunately, limited research and development funds would keep the project dormant another ten years.

By 1983, the Air Force finally faced the fact that the NT-33A had to be replaced. The Wright Laboratory was tasked to run the program and began studies that culminated in 1985 with the selection of the F-16. A conceptual design was performed to confirm that the needed modifications could be made and that the aircraft could perform the needed mission. The Air Force's Tactical Air Command "donated" an airframe not yet built so that VISTA could be built from "scratch" right on the F-16 assembly line. The prime design and production contract was awarded in 1986 to General Dynamics Fort Worth Division (which was acquired by Lockheed in 1992), with the Calspan Corporation of Buffalo, New York as the sub-contractor to design the variable stability system.

The airframe design was a mix of different features already available from other production versions, including a large dorsal housing, heavy weight landing gear, and digital flight control computer. The gun, ammunition drum, and many unneeded defensive systems were deleted. A larger capacity hydraulic pump and larger hydraulic lines were installed. A programmable center stick controlled by its own digital computer was installed in the front seat. The variable stability system design centered around three Rolm Hawk digital computers to determine the motions of the simulated aircraft and the necessary motions of the VISTA control surfaces to produce them. Controls to access the computer, in order to change flight characteristics, and to engage the front seat controls, were installed in the rear seat. Extensive automatic safety monitoring would watch the VISTA's motions and instantly return control to the safety pilot in the back seat if a potentially dangerous situation was being approached.

The Air Force approved the final design in 1988 and production began. However, as is often the case with research and devel-

One of the most complicated modifications ever made to a simulator aircraft was the VISTA modification made to this F-16. (USAF Photo)

PART I: IN-FLIGHT SIMULATORS

Prior to its completion as in-flight simulator, the same aircraft was modified with additional maneuver capabilities in this striking MATV configuration. (USAF Photo)

opment programs, limited funds resulted in a scaled down capability. While many VISTA capabilities far exceed those of the NT-33A, many other desired features had to be left out. Keeping these deleted systems in mind, the design was performed such that they could be added later as specific needs and funding became available.

The VISTA made its first flight in April 1992. The flight was scheduled for late morning, but was delayed until late afternoon. Despite the fact that the aircraft was ready, resplendent in its eye-catching red, white, and blue paint scheme, the final flight release was delayed because the paint shop had never signed off that the aircraft had been painted! After five checkout flights at the Lockheed factory at Fort Worth, Texas, the program was hit with its first major setback. Funds for the remainder of the year and for most of the next year were withdrawn. This was directed from a higher headquarters, so the Wright Laboratory had no choice but to put the program on hold.

But, as fate would have it, another Wright Laboratory program was gearing up which needed an F-16D for its test vehicle for the identical time period that VISTA would be on hold. General Electric had already developed a nozzle that could deflect the engine thrust up to seventeen degrees off the thrust axis in any direction (this could produce up to 4,000 pounds of force up, down, or sideways on the tail of the aircraft). It was anticipated that the combination of thrust vectoring and control surface motions could greatly improve the F-16's maneuverability. This program, called MATV, for Multi-Axis Thrust Vectoring, was to demonstrate the improved combat effectiveness of an F-16 equipped with this nozzle. Thus the Wright Laboratory "loaned" the VISTA aircraft to the MATV program.

Actual modifications to the VISTA aircraft began in Summer 1992. From the outside, the aircraft looked about the same, except for the addition of a spin chute and the painting of the MATV logo on the tail. The vectoring nozzle was undistinguishable from a standard one except to the trained eye. Internally, the modifications were more extensive, with most of the VISTA-unique systems removed so that from a functional standpoint, the airframe was essentially a standard F-16D. The software in the digital flight control computer was modified so that when the thrust vectoring mode was selected, the nozzle moved with the elevator and rudder. No special commands by the pilot were needed.

The VISTA made its first MATV flight in July 1993 and continued through March 1994 when it made its 95th and final MATV flight. The first phase of these flights investigated using thrust vectoring to maneuver the aircraft in the "post stall" flight regime. What this means is that the aircraft is actively maneuvered even through the wing is stalled! Traditionally, aircraft are maneuvered up to the stall, but not beyond because of the possibility of entering a spin. Using thrust vectoring to control the pitch and yaw, the aircraft's nose can still be pointed as desired, even though the wing is stalled and the control surfaces may be ineffective. Thrust vectoring allows the pilot to point the nose wherever he wants without worrying about losing control.

Sounds great in theory, but the combat effectiveness of this capability had to be demonstrated on an aircraft representative of a top-line fighter. Four specific goals of the MATV program were to

ONE-OF-A-KIND RESEARCH AIRCRAFT

With smoke pouring from its smoke generators, the VISTA/ MATV aircraft demonstrates its high angle-of-attack capability. With the nozzle deflecting the thrust upward, the high angle-of attack ability can be accomplished. (USAF Photo)

PART I: IN-FLIGHT SIMULATORS

1) determine the flight envelope for thrust-vectoring-maneuvers, 2) evaluate the close-in, combat utility of this enhanced maneuvering capability, 3) assess the aircraft's flying qualities while using thrust vectoring, and 4) examine the engine's operability at extreme attitudes and angular rates.

During the MATV program flights, the VISTA demonstrated four post-stall maneuvers. They were

- The "cobra", a fast pitch-up to vertical, or beyond vertical, then back to level flight at almost the same starting altitude. This maneuver could be used to get a quick shot at an overhead aircraft or even a missile shot at an aircraft approaching from behind.

- The "helicopter", a high yaw rate, flat, descending spiral. It looks like a flat spin, but is completely controllable. This could give the pilot a chance to take a quick 360 degree look around and even fire off weapons.

- The "J-turn", a rapid pull up to vertical, followed by a yaw until the nose is pointed down, then a sharp pull up, resulting in a very quick 180 degree heading change.

- The "hammer head", in which the aircraft starts a loop. When vertical, the pilot makes a 180 degree post stall rotation about the pitch axis, resulting in a vertical, nose down position. From there, the pilot completes the loop.

The MATV program demonstrated a new flight regime for controlled, spin-free flight. The VISTA aircraft demonstrated stabilized flight at 85 degrees angle of attack, pitch and yaw rates up to 50 degrees per second at low airspeeds where the elevator and rudder were ineffective, and near-flawless engine operation at all attitudes and angular rates.

Later in the program, the enhancement of standard combat maneuvers with thrust vectoring was studied. In performing 175 mock aerial engagements, the VISTA demonstrated a significant combat advantage over standard F-16s, both in one-on-one and two-on-two engagements. Thrust vectoring showed it could make

The MATV engine is shown in a test stand demonstrating the extent to which the exhaust can be deflected. (General Electric Photo)

an offensive aircraft more lethal and give a defensive aircraft a better chance to survive the initial encounter and then "turn the tables" and go on the attack.

With completion of the MATV program in May 1994, the VISTA/NF-16D was restored to its in-flight simulator configuration. Development flights were performed at Edwards AFB over the next seven months to check out the operation of the new computer system. These flights verified that VISTA could perform everything it was designed to perform. This phase was completed in January 1995. The VISTA then flew to Wright-Patterson AFB in Ohio where it was accepted by the Air Force.

The VISTA/NF-16D is now based at Buffalo International Airport where it is operated for the Air Force by the Calspan Corporation. It already has become a common sight in the skies over western New York state.

ONE-OF-A-KIND RESEARCH AIRCRAFT

Variable Stability B-26 In-flight Simulators

AIRCRAFT TYPE: Douglas B-26B
MISSION: Flying Qualities Research/Test Pilot Training
OPERATING PERIOD: 1951-1981
RESPONSIBLE AGENCY: USAF Aeronautical Research Laboratory
PRIMARY CONTRACTOR: Cornell Aeronautical Laboratory/Calspan Corporation

THE STORY

The development of the variable stability B-26s resulted from experiences with high-speed flight made during and after World War II. As aircraft entered transonic flight, they displayed abrupt changes in their longitudinal stability characteristics. Namely, pitch oscillations tended to last longer, not dying out as quickly as at lower speeds. Most attempts by the pilot to control the resulting motion, especially in turbulence, usually resulted in over control. By the late 1940s, researchers at Cornell Aeronautical Laboratory (CAL) used mathematical analysis techniques to begin understanding this phenomena, and recognized the need for a special variable stability aircraft to perform further research.

In 1951, the Air Force's Aeronautical Research Laboratory (a predecessor of today's Wright Laboratory), recognized the merit of CAL's ideas and agreed to sponsor the development of such a research vehicle. In fact, it was decided to develop two different aircraft: a B-26 and an F-94. The variable stability system that CAL developed for the B-26 controlled only the elevator. The aileron and rudder remained standard. Thus, only the pitch motions of the aircraft could be modified. The evaluation pilot sat in the right seat; these controls were modified with artificial feel in pitch to duplicate any desired characteristics. The safety pilot sat in the left seat, and these controls remained standard.

The following pitch characteristics of the aircraft could be varied over a wide range of values:

- Static longitudinal stability (the tendency of the aircraft to return to its original attitude after being disturbed).

- Short period frequency and damping (a quick oscillation, typically only a few seconds long and dying out quickly).

- Long period frequency and damping (a slow oscillation, typically a minute or more long, dying very slowly).

- Stick force gradients.

- Stick-to-elevator gear ratio.

The difficulty with investigating many types of problems is that as the pilot maneuvers and the aircraft changes flight conditions, most aerodynamic characteristics change. This makes it difficult at best, if not altogether impossible, for a pilot to really understand exactly what is causing a particular problem. Only a variable stability aircraft could perform the type of research needed. For the first time, the B-26 could allow researchers to change only one characteristic

One of the variable stability B-26s as it appeared in later life in Calspan markings. This particular aircraft crashed at Edwards Air Force Base in the early 1980s. (Calspan Photo)

PART I: IN-FLIGHT SIMULATORS

at a time in a systematic manner. This would allow the problems of transonic flight to be investigated and allow designers to determine ways to maintain desirable flight characteristics.

CAL began flying the variable stability B-26 as a research aircraft in 1952. The results of early tests sponsored by the Air Force showed the importance of short period frequency and damping on a pilot's ability to maintain control. This research showed there was a range of acceptable frequencies, and that the optimal one depended on parameters such as airspeed, wing loading, and the attempted maneuver.

The B-26 performed other valuable research throughout the 1950s. By 1959, funding for research in flying qualities had declined. The Air Force decided to terminate the program, but CAL felt there was still much valuable research the aircraft could perform. The Air Force eventually donated the B-26 to CAL along with two other standard ones for other uses and for spare parts. CAL immediately began finding their own research customers.

One of the early research sponsors on CAL's new B-26 was the U.S. Navy. On one such program in 1960 that was being performed at the Navy's flight test center at Patuxent River, Maryland, a fateful meeting occurred between the CAL pilot and staff members of the Naval Test Pilot School. They hit on the idea of using the B-26 to teach handling qualities to new test pilots. Since the B-26 allowed test pilots to sort out the causes of vague differences in flight characteristics, it certainly could enhance the training of new test pilots by bridging the gap between theory and practice. Coupled with classroom lectures to explain the scientific principles and review the mathematical equations that explain the phenomena, the instruction flights could give a real "hands-on" feel and make the unexciting equations come to life.

CAL pilots and engineers, working with the Naval Test Pilot School, laid out a curriculum of classroom instruction and training flights that complimented text book work. The program started on a trial basis in 1960 and was an immediate success. So successful, in fact, that the Air Force Test Pilot School wanted to incorporate the same curriculum.

The B-26 could not accommodate the additional work, so CAL decided to convert one of the other B-26s donated by the Air Force into a second variable stability aircraft. It was also decided to expand the variable stability capability to include the roll and yaw axes in both aircraft (the third B-26 was cannibalized for a supply of spare parts). The second B-26 was completed in early 1963 and began supporting the Air Force Test Pilot School in February 1963.

The B-26s continued performing research programs and supporting the two test pilot schools throughout the 1960s and 1970s. In this respect, they started a legacy for test pilot training that includes the NT-33A, Calspan Learjets, NF-16D, and ASTRA Hawk, that continues to this day. In the late 1970s, it was realized that the B-26s could not last forever. Calspan (as CAL was now known) began developing a variable stability Learjet to replace both B-26s.

One of the B-26s was lost in a tragic crash at Edwards Air Force Base in spring 1981 (author Steve Markman flew this B-26 barely one year earlier). Two student test pilots and the Calspan pilot were killed, to date the only known fatalities involving a variable stability aircraft. It was ironic that the aircraft was on its last deployment to the Air Force Test Pilot School before being retired. The cause of the crash was determined to be a failure of the main wing spar, a problem that had caused the crash of other B-26s.

The surviving B-26 sat inactive at Calspan's hangar at Buffalo, New York, for several years. After such a long and successful career, there were emotional arguments both for and against returning it to service. Eventually it was decided to donate the aircraft back to the Air Force. It was ferried to Edwards Air Force Base in November 1986 and placed in storage for eventual display at Edward's flight test museum.

ONE-OF-A-KIND RESEARCH AIRCRAFT

NC-131H Total In-Flight Simulator

AIRCRAFT TYPE: C-131 "Samaritan" Military Transport
MISSION: In-Flight Simulation, Research
OPERATING PERIOD: 1970 - Present
RESPONSIBLE AGENCY: Wright Laboratory, Wright-Patterson Air Force Base, Ohio
PRIMARY CONTRACTOR: Calspan Corporation

THE STORY
The U.S. Air Force's NC-131H Total In-Flight Simulator (or TIFS, for short) has a long history of performing research and in-flight simulations, exceeded only by the history of the NT-33A. As with the NT-33A, the TIFS was developed by Cornell Aeronautical Laboratory (now Calspan Corp, Buffalo New York), and now operated by Calspan for the Wright Laboratory at Wright-Patterson AFB, near Dayton, Ohio.

The TIFS concept evolved over several years, starting in the late 1950s. When it was finished in the early 1970s, the TIFS was the most advanced in-flight simulator ever built. After over twenty years of operation and numerous upgrades to its systems, it continues to be the most sophisticated and capable in-flight simulator in the world.

The development and selling of the original TIFS concept is an exciting story by itself. It shows how the persistence of a group of far-sighted engineers at Cornell Aeronautical Laboratory, coupled with the needs of industry and Government, utilized the best resources and capabilities of each partner to turn an idea into reality. The earliest ideas for TIFS were as a training aircraft for pilots learning to fly the first generation of jet airliners. The first sketches showed a Convair prop airliner with a jet cockpit on top of the fuselage. The flight characteristics of the new jets would be modeled on computers. When the pilot in the jet cockpit moved the controls, they would be directed to the computer, which would then move the control surfaces so that the in-flight simulator's response would exactly match that of the aircraft modeled in the computer. In addition to matching the flight characteristics, the total cockpit environment would be duplicated.

This idea seemed a natural for the airlines, who were expected to jump at the opportunity to help develop the aircraft. The safety implications and long-term economics of the idea were obvious. However, several years of marketing the idea to the airlines resulted in no significant interest.

By the summer of 1963, the Federal Aviation Administration (FAA) was actively pursuing development of the Supersonic Transport (SST). Cornell engineers held several meetings with the FAA at this time, marketing the idea not so much as an airline trainer, but as a development tool for the SST. The use of variable stability aircraft for research and development of new aircraft was becoming common, and the proposed in-flight simulator could be designed around the research needs of the SST. The name "Total In-Flight Simulator", or TIFS, for short, was born at one of these meetings with the FAA.

By the fall of 1963, similar discussions were held with the Air Force Flight Dynamics Laboratory (a predecessor of the current

The C-131H TIFS in-flight simulator is one of the most recognizable of the production-turned-research aircraft. The C-131 is highly instrumented and carries a low-mounted front nose. (NASA Photo)

PART I: IN-FLIGHT SIMULATORS

Wright Laboratory) at Wright-Patterson AFB, Ohio. With programs like the C-5 transport and B-1 bomber in early conceptual stages, there was interest in the TIFS, but research money was not available to sponsor development. Cornell continued pitching the TIFS through the mid-1960s as a possible joint development by the FAA and the airlines to support SST development.

By this time, the TIFS concept had evolved to an in-flight simulator with a separate "evaluation" cockpit on the nose. Several noses could be built and interchanged quickly as different aircraft were simulated. The TIFS would be able to duplicate the simulated aircraft motions in all six degrees of freedom. To do this, hydraulic actuators would be added so the computer could move the elevator, aileron, and rudder. To produce linear motions, the flap would be modified into a "direct lift flap", capable of moving up and down at a high rate to control lift. Vertical fins, or side force surfaces, would be added on the wings to control the side acceleration. A servo control would be added to the prop governor to make the TIFS speed up or slow down. The motions of all these control surfaces would be orchestrated by a specially designed analog computer system that would monitor the evaluation pilot's control inputs, determine the required aircraft response, and then move the control surfaces to produce that response.

Cornell's persistence finally paid off in April 1966 when the FAA, Flight Dynamics Lab, and the C-5 program office agreed to pool funds to sponsor development of the TIFS. With the Air Force being the major sponsor, the development would be managed by the Flight Dynamics Laboratory. The development contract was finalized in December 1966, and the Air Force delivered a C-131B (the military version of the Convair 340) to Cornell's facility at Buffalo, New York.

Design and construction started promptly. Much of the nose structure was modified to support the evaluation cockpit and other modifications to the wing were made. In December 1968, the TIFS was ferried to Burbank California where the original piston engines were replaced with Allison 501-D13 turboprop engines. This commercially-available modification nearly doubled the available horsepower and brought the basic airframe configuration up to the equivalent of the Convair 580, with the military designation C-131H.

Other modifications followed in 1969 and the newly-completed TIFS made its first flight in June 1970. Nearly a year of checkout flights were required to verify operation of all the specially designed systems and to confirm the capabilities of the new aircraft, officially designated NC-131H (the "N" meaning that the aircraft was permanently modified and never to be returned to its original configuration).

The first research project was a simulation of the B-1 in June 1971, to verify the anticipated flight characteristics of the new strategic bomber. Later, the TIFS was ferried to the Northrop factory at Palmdale, California, where test pilots flew it over the actual route the B-1 prototype would fly on the maiden flight from Palmdale to nearby Edwards AFB. Following the B-1's first flight, the test pilots reported that "It flew just like the in-flight simulator . . .", one of the highest compliments they could have made.

Other projects followed rapidly in the first few years. NASA sponsored a short simulation of the Space Shuttle in March 1972. This was followed in April-July by a simulation of the Concorde

TIFS aircraft on the ground at Kelly Air Force Base in 1985. (USAF Photo)

ONE-OF-A-KIND RESEARCH AIRCRAFT

SST. Although the American SST had already been canceled, the British and French consortium building the Concorde wanted to certify it in accordance with FAA criteria. However, the FAA standards lacked data pertinent to the Concorde's unique characteristics. Thus, the FAA sponsored this simulation to collect data and determine acceptable and unacceptable flight characteristics, especially for landing with failures in the Concorde's stability augmentation.

The first of many "flying laboratory" programs was flown in 1973 when a study of cross-wind landings was performed. The side force surfaces were used to demonstrate the value of direct side force control to cancel the effects of cross winds. Two modes were mechanized. In the first, the pilot in the simulation cockpit manually deflected the side force surfaces by means on a thumb switch mounted on the control wheel to cancel the effect of the cross wind. In the other mode, an automatic control system mechanized on the TIFS' computers deflected the side force surfaces in order to track the localizer, allowing the pilot to make a wings-level, no crab, landing in direct cross winds of up to fifteen knots.

Throughout 1974 and 1975, NASA again sponsored several additional simulations of the Space Shuttle, especially looking at the final steep approach, the roundout, and the flare and simulated touchdown. The windscreen in the simulation cockpit was masked to produce the proper out-of-the-cockpit view, and a control stick similar to the Shuttle's was installed. Duplicating the steep final approach that exceeded both two hundred fifty knots of airspeed and a fifteen degree glideslope required a lot of extra drag. This was done by lowering the landing gear and deflecting the side force surfaces opposite to each other. Simulated touchdowns were made in which the TIFS landing gears did not actually touch the runway. This was because of the high speed at which the Shuttle would touch down and because the pilot in the Shuttle sat much higher than in TIFS, producing very different visual cues. When the TIFS' computer determined that the Shuttle's main gear should touch the runway, it sent a quick abrupt signal to the direct lift flap to produce a "bump", indicating that the shuttle had landed. Although the TIFS was still several feet in the air, the pilot was at the correct height above the runway, resulting in the correct visual scene. Other Shuttle simulations were performed in 1978 and 1985 to help improve Shuttle flight characteristics.

In 1976, in a program sponsored by the Flight Dynamics Laboratory, the TIFS helped develop an autoland system for a new Remotely Piloted Vehicle (RPV) being designed. In this program, the flight characteristics of the Ryan YQM-98A and the autoland system were modeled on the TIFS computer. The test program demonstrated the ability of the system to track an instrument landing system and land the aircraft safely, including deceleration to a stop on the runway. In addition, the feasibility of flying the YQM-98A manually from a ground station was demonstrated. The pilot sat at a control console that had normal flight instruments and flight controls and displayed a television picture sent from a camera mounted in the TIFS' nose. This program was flown at Wright-Patterson AFB, bringing TIFS "home" for the first time.

Another "flying laboratory" program was flown in 1978, this time to gather data needed to design better ground-based flight simulators. Motion cues are very important in a flight simulator, and must be harmonized properly with the visual scene presented to the pilot. Ground-based simulators have one obvious drawback - they have very limited distances over which they can travel. An initial motion cue must be given to the pilot, then very carefully washed out before the simulator reaches the full extent of its travel. Improperly designed motion washouts can result in the simulator producing an improper feel. In this experiment, different tests were performed to determine how the subject pilots perceived different motions. The subject pilots had no outside vision, and flew the TIFS using instruments while different motion effects were used. Once again, the TIFS was the only aircraft that could have performed such research because of the ability to control all aspects of motion, from none at all up to full and sustained motions.

In 1980, another "generic" flying qualities study was performed to gather data to develop a flying quality specification for very large aircraft. Over the years, the military and airlines developed jumbo jets such as the C-5 and Boeing 747, and the trend was toward larger, more massive aircraft. However, there was little data defining desirable flight characteristics for an aircraft of this size. The problem with flying such large aircraft is that the pilot is sitting far ahead of the center of rotation. Even small bank or pitch motions could throw the pilot through large motions that would make controlling the aircraft difficult. The TIFS' computers were programmed with the predicated flight characteristics of several generic one million pound aircraft. The aircraft motions were evaluated by test pilots performing a variety of flying tasks.

Developing futuristic cockpit displays to make complicated flight missions easier through the use of graphical displays has long been a goal of human factors engineers. Many promising displays were developed using ground-based simulators. The problem with most was that the size and weight of the computer needed to perform all the computations and generate the picture precluded their being installed in an aircraft. In 1983 such a display, called the "Command Flight Path", was programmed on a computer small enough to fit in TIFS. The display essentially drew a path between navigation fixes, thus giving the pilot a pictorial of the route he was to fly. An aircraft symbol, representing his aircraft, was shown several hundred feet ahead. All the pilot had to do was fly a loose formation with the aircraft on the display. This test program showed by actual flight testing the effectiveness of this type of display, and paved the way for many cockpit displays used today.

Throughout the 1980s, many other programs were flown, including development support for the X-29 research aircraft, B-2 stealth bomber, and YF-23 tactical fighter prototype. Other research programs were also flown supporting research in flying qualities and flight control system designs, and evaluating protective drugs for pilots to take prior to entering chemical warfare areas. By the mid-1980s, the Air Force began making the TIFS available to support commercial developments. Simulations were performed of an advanced Boeing jet aircraft, called the 7J7 (which was eventually canceled), and the Douglas MD-12, a new wide-body airliner currently in preliminary design.

Also in the mid-1980s, another unique use for TIFS was started. The Air Force and Navy test pilot schools funded the development of the Avionics System Test Training Aircraft (ASTTA) capability for the TIFS. Most test pilots never get the opportunity to fly a new prototype aircraft on its first flight. Rather, they spend their testing careers verifying capabilities of aircraft and testing new systems developed for existing aircraft. The test pilot schools developed this system as a tool to train new test pilots (who were already experienced pilots) to test avionics systems. The simulation nose was removed and replaced with a large radome. Installed inside the radome was an F-16 radar unit; under the nose a forward looking infrared (FLIR) sensor; internally, a forward looking television cam-

PART I: IN-FLIGHT SIMULATORS

The TIFS' simulation nose certainly makes this a unique one-of-a-kind aircraft. (Calspan Photo)

With radome installed, TIFS is used to teach avionics testing. (Calspan Photo)

era, an inertial navigation system, a student work station from which to operate the systems and view the displays, and a computer system that could read and record internal signals from each of the avionics pieces. Over the years several additional systems were installed, including the seeker head from a television-guided Maverick missile, a Global Positioning System receiver and displays, and a LORAN navigation system. The TIFS can be converted between the simulation and ASTTA configurations in only a few days. Twice each year, the TIFS travels to Edwards Air Force Base and Patuxent River Naval Air Station to help train new test pilots to perform avionics testing.

Other system upgrades over the years have increased TIFS' capabilities and supportability beyond its original design. Most of the analog circuitry in the variable stability system has been replaced with modern airborne quality digital computers, and the recording system sports state-of-the-art digital data recorders. A color multi-function display, Head-Up Display, and other high-resolution graphics systems in the evaluation cockpit can be programmed as needed for individual simulation programs. A variety of control sticks and control wheels that were developed over the years for other programs are kept at hand and can be reinstalled as needed. The engines have been replaced with more powerful Allison 501-D22G engines and the original Aeroproducts props replaced with Hamilton Standard props, identical to those on the latest C-130s.

Although the TIFS airframe is forty years old, it remains a valuable research tool with no plans for retirement. Although the Air Force has retired all other C-131s and provides no depot support, many Convair airliners remain in commercial use throughout the world. The TIFS will likely remain operational as long as commercial support is available because the airframe, unique airframe modifications, and basic design of the variable stability system still provide a viable capability for future needs. The Air Force is studying system upgrades and structural improvements that will keep the Total In-Flight Simulator performing valuable research well into the next century. The longevity and variety of uses never envisioned in the TIFS' original design stand as a tribute to the visions of forward-looking people who conceived the TIFS over thirty years ago.

ONE-OF-A-KIND RESEARCH AIRCRAFT

Gulfstream Shuttle Training Aircraft

AIRCRAFT TYPE: Grumman Gulfstream II
MISSION: Space Shuttle Pilot Training
OPERATING PERIOD: 1976 - Present
RESPONSIBLE AGENCY: NASA
PRIMARY CONTRACTOR: Grumman

THE STORY

One of the greatest honors and challenges a pilot can receive is being selected to fly the Space Shuttle Orbiter. It is also one of the toughest assignments. The Space Shuttle is not an easy aircraft to fly. It is the world's biggest, heaviest, and fastest glider. It has to be landed on the first try by a pilot who has become unadjusted to the effects of gravity. When the Shuttle was being developed, training the pilots was recognized as one of the major challenges.

The solution was to develop an in-flight simulator that would fly just like the Space Shuttle. Thus was born the Shuttle Training Aircraft, or simply STA. The STAs are modified Grumman Gulfstream II executive aircraft. The astronaut-pilot flies from the left seat with a safety pilot seated next to him. Four STAs are operated by NASA. Two were delivered in 1976, a third in 1985, and a fourth in 1990.

The Gulfstream II was selected for the STA mission because its speed and altitude capabilities allowed it to match the Shuttle's final approach from about 35,000 feet until landing. The Gulfstream II also has a large cabin that can house computer and recording equipment, a spacious flight deck, and a large fuel capacity so that ten or more climbs to 35,000 ft can be performed on a single mission.

Major modifications included:

• Replacement of the left pilot's instrument panel and flight controls to duplicate those of the Shuttle

• Installation of a model-following flight control system in which the aerodynamic characteristics of the Shuttle are modeled

• Addition of a simulation engineer's work station to operate the computer system and recording equipment

• Replacement of the Fowler flap with a fast-moving simple flap to provide direct lift control (this allows simultaneous matching of lift characteristics and over-the-nose visibility)

• Removal of the spoilers and mechanizing differential flaps to enhance roll control (this was done because the spoilers, normally used to augment the ailerons, had an adverse affect on the direct lift control)

• Addition of vortex generators and wing fences to improve the direct lift control

Shuttle Training Aircraft performing a shuttle-type steep approach. Note the side-force generator located under the fuselage which was eventually removed. (NASA Photo)

PART I: IN-FLIGHT SIMULATORS

- Installation of fuel tank baffles to prevent fuel from shifting within the fuel tanks and causing the center of gravity to shift forward during steep approaches

- Reinforcement to the landing gear to allow extension at high speeds to produce extra drag needed for the steep approach

- Modification of the thrust reversers to allow them to operate in flight, to produce additional drag

- Structural reinforcements to the engine area to accept increased reverse thrust loads

A typical STA mission profile starts with a climb to 35,000 feet. Once at the starting altitude, the simulation system is engaged so that the pilot-astronaut is flying. He then intercepts the navigation signal from the instrument landing system, flies the steep approach that is virtually identical to the Shuttle's, and finally performs a simulated landing. Because the Shuttle is much longer, the STA is still several feet in the air when the Shuttle would contact the runway. Thus the STA's wheels do not actually touch the ground. These proper-eyeheight "landings" are performed because the pilot has to be at the same height above the runway as he would be in the Shuttle to have the correct visual scene. Ten to twelve such approaches are flown on each mission.

The STAs are used in three different ways to train pilot astronauts: basic training, proficiency training, and mission-specific training. During basic Shuttle training, the new astronaut-pilot becomes familiar with the Shuttle's controls and displays, flying the steep approach, Shuttle handling qualities, effects of variation in weight and center of gravity, and proper corrective responses to off-nominal conditions. Each astronaut-pilot flies twenty STA flights over a six month period while in basic training.

Once completing basic training and awaiting assignment to a specific flight, each astronaut-pilot receives two STA flights per month to maintain flight proficiency.

Mission-specific training is performed once the astronaut-pilot receives a mission assignment. In this phase, approaches under conditions that are relevant to that flight are flown. These conditions include day or night approaches as appropriate with the flight characteristics adjusted for the actual weight and center of gravity for the mission. Practice approaches are performed at both the primary and backup landing sites. About ten flights are flown over a nine-month period to prepare the astronaut-pilot for the specific mission.

By the time an astronaut is ready for his first flight in the shuttle, he will have performed about 500 approaches and simulated landings in one of the STAs.

The STAs are also used for various types of mission support, such as flying approaches at the landing site just prior to an actual Shuttle landing to evaluate atmospheric conditions that the Shuttle will encounter.

The STAs fly an exhausting schedule, about 200 missions per year for about 500 to 600 flight hours each. When not flying, they can be connected to ground-based computers and "flown" on the ground as a fixed-base flight simulator.

ONE-OF-A-KIND RESEARCH AIRCRAFT

ASTRA Hawk In-Flight Simulator

AIRCRAFT TYPE: British Aerospace Hawk Mk 1
MISSION: In-Flight Simulation/Test Pilot Training
OPERATING PERIOD: 1986 - Present
RESPONSIBLE AGENCY: Empire Test Pilot School, Royal Air Force
PRIMARY CONTRACTOR: Cranfield Institute

THE STORY

Using in-flight simulators to train new test pilots has been well-established in the United States and England for many years. Just as the U.S. Air Force and U.S. Navy test pilot schools use the Calspan Learjets and USAF NT-33A, the Empire Test Pilot School (ETPS), located at Boscomb Down near Salsbury, England, also uses this unique capability to train new test pilots for the Royal Air Force. The Advanced Systems, Training and Research Aircraft, or ASTRA Hawk, serves this role. The ASTRA Hawk is the only in-flight simulator owned and operated exclusively by a test pilot school.

The ASTRA Hawk is used in the ETPS syllabus to train new test pilots. Flight characteristics can be changed quickly to make the aircraft responsive, sluggish, or anywhere in between. Stick characteristics and head-up display formats can also be changed. The ASTRA Hawk is supplemented by an older, variable stability Basset, which is a twin engine general aviation aircraft. The Bassett has been in service for over twenty years and flies at speeds of 100-150 knots.

The selection of the BAE Hawk trainer for the ASTRA mission resulted from studies performed by the Cranfield College of Aeronautics in the early 1980s. Cranfield then was awarded the follow-on contract for the design, modification, and flight testing. The Hawk was chosen for several reasons: 1) it was currently in the RAF inventory and would remain so for the foreseeable future, 2) its flight envelope was representative of modern tactical combat aircraft, 3) the ETPS already had three other Hawks in its fleet, and 4) early studies showed that the Hawk could be modified to perform this new role. Limitations in cockpit size, rudder design, and some stability characteristics in the basic Hawk configuration were overcome to make the ASTRA Hawk a very successful and versatile training tool.

In being converted into an in-flight simulator, the ASTRA Hawk's front controls were disconnected from the flight control system. Other pilot-in-command functions were also moved to the rear seat to allow solo flight to be performed by the safety pilot, if necessary. Rotary hydraulic actuators were connected to the push rods of all three control surfaces so that the computer could move each one as needed. The front cockpit features a HUD, center stick, side stick, and rudder pedals. Except for the side stick, all characteristics can be programmed and changed in flight. The side stick cannot be adjusted in flight as on many other in-flight simulators, but can be adjusted on the ground to produce the stick characteris-

Empire Test Pilot School's variable stability ASTRA Hawk, with the older variable stability Basset in the background. The number "1" on the vertical tail is just about the only external feature distinguishing this aircraft from the school's other Hawk aircraft.

tics needed for any particular flight. The rear cockpit has all controls needed to engage, disengage, and monitor the operation of the variable stability system.

With the variable stability system engaged, the ASTRA Hawk can obtain up to 14 degrees angle of attack (which coincides with stall buffet onset), speed as great as .7 Mach, normal acceleration from -1.75g to +6.5g, roll rates up to 200 degrees per second, and pitch rates up to 40 degrees per second. This envelope allows flight test techniques to be taught for maneuvers representative of most tactical aircraft.

The ASTRA Hawk was completed and made its first flight in 1986, but did not become operational in the ETPS curriculum until 1990. Rigid British concerns for safety precluded actual landings while flying from the front seat. However, as ETPS gains experience with the aircraft and the regulatory agency builds confidence in its reliability, this limitation should eventually be lifted.

It is interesting to note the difference in American and British philosophies regarding reliability of the electronic systems versus human operators. The computer systems used on in-flight simulators are typically single channel. This means that only one set of computations are performed and if the system fails, it disengages and the safety pilot must take over. American in-flight simulators routinely operate to touch down because the safety pilot is considered the ultimate authority. The British philosophy is that the safety pilot may not be able to respond quickly enough close to the ground, and thus the computer system must demonstrate that it is reliable enough to remain in control.

Students at ETPS currently receive at least four training flights in the ASTRA Hawk, but future students may receive more as added roles for the aircraft are found. In a typical year, the aircraft flies about one hundred hours and is used to train about fifteen new test pilots.

PART I: IN-FLIGHT SIMULATORS

University Of Tennessee Navion In-Flight Simulators

AIRCRAFT TYPE: North American Navion
MISSION: Flight Control Research
OPERATING PERIOD: 1960s - Present
RESPONSIBLE AGENCY: Princeton University, University of Tennessee Space Institute (UTSI)

THE STORY

The two Navion aircraft are excellent examples of the use of general aviation aircraft for flight research. They demonstrated that valuable, military-oriented research does not always have to be performed on expensive military research aircraft. Simpler, economical aircraft often can do the job. The important factor is the suitability of the vehicle for researching the problem of concern.

The Navions were developed by Princeton University Flight Laboratories in the 1960s, and operated by them through the early 1980s. Both are late 1940s vintage aircraft, one built by North American and the other by Ryan. They were modified under a Naval Air Systems Command (NASC) contract, and performed a variety of research programs for NASC, the Air Force, and NASA. Some of the research projects performed on these aircraft included investigation of:

- Sideforce control and controllers
- STOL approaches
- Powered lift longitudinal handling qualities
- Lateral-directional aircraft dynamics

The two Navions currently are registered as N55UT and N66UT. These roomy, four-place aircraft feature the following modifications:

- The addition of hydraulic actuators to the elevator, aileron, and rudder, a hydraulic pump capable of running all actuators at full rate simultaneously, and all the plumbing needed to connect the pump to all actuators

- The modification of one set of controls for use by the evaluation pilot, while the other set of controls for the safety pilot is essentially standard

- Installation of digital and analog computers to perform all flight control and data processing computations

- Structural reinforcement to the landing gear to allow large sink rates at touchdown

- Gross weight increase, because the empty weight is near the maximum gross weight for an unmodified Navion

- Modification of the flaps into direct lift flaps that move 30 degrees up and down (as compared to 40 degrees down for the standard flap). They can double as wing spoilers for extra drag either in flight or on the ground. The fast acting actuators can drive the direct lift flaps at up to 110 degrees per second in either direction

- A data acquisition system that can monitor and telemeter up to 43 channels, including angular rates, attitudes, accelerations, heading, control inputs, performance measures, altitude, glideslope data, and main strut compression

- An automatic abort mode, that disconnects the evaluation pilot's controls, advances the power to 75 percent, and resets the flaps to 20 degrees when the safety pilot commands a disengage. During a normal approach, less than 10 feet of altitude is lost after initiating an abort

- Dual channel fly-by-wire control of the aileron and elevator, and single channel control of the rudder, flaps, and throttle

N66UT was built in 1949 by North American. The evaluation pilot sits on the left and the safety pilot on the right. The landing gear is bolted in the down position.

N55UT was built in 1946 by Ryan. It is the more extensively modified of the two Navions. The evaluation pilot sits on the right and safety pilot sits on the left, opposite of N66UT. It also features side force surfaces mounted on the wings, giving N55UT full six degree-of-freedom capability. It has a larger hydraulic pump so that all control surfaces and the landing gear can operate at the same time. The vertical tail and rudder are larger than standard to compensate for the loss of directional stability caused by the side force surfaces.

By the early 1980s, the research projects for the Navions decreased sharply. After sitting dormant for about five years, Princeton University sold the aircraft to University of Tennessee Space Institute in Tullahoma, Tennessee. UTSI restored the aircraft to full capability, and uses them in graduate engineering courses and in other specialty short courses. They remain available for a wide range of aviation research and specialized training applications.

Left: UTSI Navion on the ground. Note the wing-mounted angle-of-attack and side slip sensors. Right: The UTSI Navion in flight. Note the tuffs of yarn on the starboard side-force surface to study airflow. (Princeton University Photos)

ONE-OF-A-KIND RESEARCH AIRCRAFT

P-2 Variable Stability Aircraft

AIRCRAFT TYPE: Lockheed/Kawasaki P-2H
MISSION: In-Flight Simulation/Flight Control Research
OPERATING PERIOD: 1978 - mid 1980s
RESPONSIBLE AGENCY: Third Research Department, Japan Defense Agency
PRIMARY CONTRACTOR: Kawasaki

THE STORY

The Kawasaki P-2 Variable Stability Airplane (VSA) was developed and operated in Japan to investigate new flight control concepts for future aircraft. It was built from the 39th production P2V-7 (P-2H) Neptune, an anti-submarine patrol aircraft designed by Lockheed, but built under license in Japan by Kawasaki Heavy Industries.

As do most other larger countries, the Japanese government performs flight research. The Third Research Department of the Japan Defense Agency's Technical Research and Development Institute developed and operated the P-2 VSA as an in-flight research vehicle. The actual modifications to the aircraft were made by Kawasaki during 1977 and the aircraft made its first flight in December 1977 at Gifu Air Base.

Major modifications to the airframe included:

• Removal of all anti-submarine warfare equipment, including the radar

• Installation of a fly-by-wire flight control system while retaining the mechanical backup for the safety pilot

• Replacement of the right-seat pilot's mechanical controls with variable ones that connect to the computer system

• Installation of a computer capable of performing all flight control computations and determining forces and motions for the variable feel control wheel

• Replacement of the outboard trailing edge flaps with direct lift flaps that could move 40 degrees down and 20 degrees up and the addition of vortex generators just ahead of them to improve effectiveness

• Addition of side force surfaces on the top and bottom of each wing outboard of the auxiliary jet engine, capable of moving 30 degrees to the left or right

• Addition of two control operators' stations in the fuselage bay

• Addition of speed brakes to the inside of each weapons bay door

• Installation of flight instrumentation and recorders

As an in-flight simulator, the P-2 VSA was capable of independent control of five degrees of freedom: pitch, roll, and yaw rotations, and vertical and side translations.

During 1978 the aircraft was operated by the 51st Air Squadron of the Japanese Maritime Self Defense Force. It was used to evaluate the effectiveness of direct lift and direct side force control in the overall operation of the aircraft. Following these evaluations the P-2 VSA was used at the Japanese Maritime Self Defense Force's test pilot school to train new test pilots. It was retired during the 1980s.

P-2 Variable Stability Aircraft in flight. The aircraft performed airborne simulation and research. (Japan Defense Agency Photo)

PART I: IN-FLIGHT SIMULATORS

S-76 SHADOW Helicopter In-Flight Simulator

AIRCRAFT TYPE: Sikorsky S-76
MISSION: In-Flight Simulation/Display Research
OPERATING PERIOD: 1986 - Present
RESPONSIBLE AGENCY: Sikorsky Aircraft Co

THE STORY

The Sikorsky S-76 SHADOW was developed and operated by the Sikorsky Aircraft Company to support helicopter research and development. SHADOW stands for Sikorsky Helicopter Advanced Demonstrator of Operator Workload. It is an in-flight simulator that can duplicate a wide variety of helicopter dynamics, control techniques, instrument displays, and cockpit visibilities. The SHADOW was developed in the mid-1980s to support the U.S. Army's Light Helicopter Experimental (LHX) program, which developed into the RAH-66 Comanche. The SHADOW remained in service to support other helicopter-related research.

The LHX helicopter, whose development SHADOW was designed to support, was to be a single pilot combat helicopter that would perform most of its flight at very low altitude and often at night or in poor visibility. This type of mission requires the pilot to keep his head out of the cockpit most of the time, yet still absorb and analyze large amounts of information – a scenario that would keep any pilot extremely busy. For this type of extremely high pilot workload, ground-based simulators alone would not be an adequate development tool because of their inherent limitations of vision, motion, and "pucker factor." To develop such a new helicopter, an in-flight simulator capable of duplicating all the essential elements of helicopter flight in a realistic environment was needed.

The S-76 proved to be an excellent airframe from which to build the SHADOW. The fifth prototype was selected as the airframe to be modified. It had supported early S-76 development and had been unused for several years. The central keel-beam support structure could be extended to support the weight of the evaluation cockpit. The internal volume was adequate to allow for two safety pilots, a flight test engineer, and an observer. Last, the avionics and baggage areas were large enough to house all the computers, recording equipment, and experimental black boxes that would need to be carried. The engines were also modified to produce additional power that would allow single engine hover at mission gross weight and at moderate density altitudes.

The SHADOW features the following capabilities:

- Separate Evaluation Cockpit - The evaluation cockpit was added in front of the nose. It was designed to allow maximum

The extra cockpit allows the Shadow in-flight simulator to study single-cockpit helicopter operations. (United Technologies Photo)

ONE-OF-A-KIND RESEARCH AIRCRAFT

The Shadow in-flight simulator is shown in a typical low-altitude mission profile. (United Technologies Photo)

- Fly-by-Wire Flight Controls - The flight controls in the evaluation cockpit are electronically connected to the SHADOW's computer system. New types of cockpit controls and control mechanizations can be tested safely because they are electronically connected to the SHADOW's computers, rather than connected directly to the mechanical controls. Specific characteristics of the controls can be varied to duplicate almost any desired set of characteristics, either existing or theoretical.

- Navigation System - Navigation equipment consists of inertial and Doppler navigation systems, satellite based global positioning, radar altimeter, digital maps, and infrared vision systems. These are all integrated through the navigation computer so that various new mechanizations using any of the data can be tested.

- Simulation Capability - The helicopter being simulated is modeled in the SHADOW's computers. As the evaluation pilot moves the flight controls, the computer model determines the response of the simulated aircraft and moves the SHADOW's controls to make the SHADOW's motions match the model. The system monitors itself and notifies the safety pilots of any approaching system limits, malfunctions, or potentially dangerous situations. These capabilities are similar to those found in many other in-flight simulators.

- Safety Pilot/Flight Crew Capability - Two safety pilots fly from the conventional S-76 cockpit and monitor the flight and the evaluation pilot's performance. At any time they can take control of the aircraft either by disengaging the evaluation pilot's control or pulling on the stick hard enough to override the evaluation pilot's commands. By having two safety pilots, actual low-level night missions can be flown. The flight test engineer monitors the operation of all systems, can change models being evaluated, and can change specific parameters within the model. A forth seat can be occupied by an observer. All crew members can monitor the evaluation pilot's performance by watching the computer-generated displays and the over-the-shoulder video sent from the evaluation cockpit.

flexibility for investigating single-pilot operations during a low altitude combat mission. It has a large glass area to allow the maximum viewing area possible. For specific tests, the glass can be masked as needed to duplicate a proposed viewing area. A modular instrument panel allows flight instruments and controls to be changed or rearranged easily. The standard instrument display consists of two cathode ray tubes and one liquid crystal display. These can be replaced by mechanical flight instruments as needed. Other hardware, such as experimental control sticks can be installed quickly.

- Programmable Display System - The display processor (which is really just a specialized digital computer) performs all the computations needed to generate the electronic displays. It can be programmed easily to generate virtually any display for the evaluation pilot to test. The pictures can then be projected on either multi-function displays, liquid crystal displays, or helmet-mounted displays in the evaluation cockpit. In addition, they can be displayed elsewhere in the aircraft and monitored by the crew.

In the years since the SHADOW was developed, it has performed a variety of other research programs. These include helicopter related research in the following areas:

- Cockpit Visibility - The specific needs for out-of-the-cockpit visibility vary with the mission being performed. Plastic canopies and windows are heavy and add little strength. Thus, the less window the better. The exact amount of visibility that must be provided becomes a major design tradeoff and the design of the canopy lines is not an arbitrary decision. The lines must be carefully selected to allow the required distortion-free visibility yet keep the glass to a minimum to decrease weight.

- Helmet-Mounted Displays - This relatively new display technique projects graphical and alphanumeric information directly onto the pilot's visor. This allows him to keep his eyes completely out of the cockpit. Due to the limited amount of weight the pilot can support on his head, the equipment must be kept as small and light as possible. However, weight tends to increase in proportion to the resolution and field of view. The

PART I: IN-FLIGHT SIMULATORS

The Shadow evaluation cockpit features dual CRT displays and experimental control sticks. (United Technologies Photo)

size of the display and the amount of resolution needed to perform the mission are important design factors that are best determined from actual flight test.

• Controllers - Just as with fixed-wing aircraft, the design of the control sticks in helicopters is very critical. An aircraft with good flight characteristics can feel bad to the pilot if the stick is not correct. A good stick design can give the helicopter a good feel and made it easier to fly. Normal helicopter control require two hands and two feet: the right hand for pitch and roll control, the left hand for throttle, cyclic, and collective control, and two feet for yaw control. Studies looked at alternate mechanizations, such as a right hand four-axis controller that controls pitch and roll as normal, but twists for yaw and moves vertically for collective control. The purpose of such investigations is to look at ways to decrease pilot workload by making the controls more intuitive and natural feeling.

ONE-OF-A-KIND RESEARCH AIRCRAFT

NT-33A In-Flight Simulator

AIRCRAFT TYPE: T-33A "Shooting Star" Military Jet Trainer
MISSION: In-Flight Simulation, Research
OPERATING PERIOD: 1957 - Present
RESPONSIBLE AGENCY: Wright Laboratory, Wright-Patterson Air Force Base, Ohio
PRIMARY CONTRACTOR: Calspan Corporation

THE STORY

The NT-33A surely holds the record as being the longest continuously-operated test aircraft in history, as well as being the most productive in-flight simulator ever. Having been operated as an in-flight simulator since 1957, it has been used to support the development of aircraft from the X-15 up to the YF-22. In addition, it supported a large amount of aeronautical research and was used to train most U.S. Air Force and U.S. Navy test pilots since the mid-1970s. With the retirement of the T-33 fleet in the late 1980s, the NT-33A became the last remaining T-33 in Air Force service. With a 1951 tail number, it has been the oldest flying aircraft in the Air Force since the mid-1980s.

The NT-33A was delivered to the Air Force in October 1951, but remained at the Lockheed factory as an engine-airframe testbed. In June 1952 it went to the Allison Division of General Motors to continue engine test work. In October 1954 it was delivered to the Cornell Aeronautical Laboratory (now Calspan Corporation) in Buffalo, New York, to be re-built as an in-flight simulator. This work was to be performed under a contract from the Wright Air Development Center (now Wright Laboratory) at Wright-Patterson Air Force Base in Ohio.

The aircraft was modified extensively over the next several years. The major modifications included:

• Replacement of the original nose with the larger nose from an F-94 aircraft (the F-94 was an interceptor based on the T-33 airframe). The larger nose was needed to house all the recording equipment and analog computers needed to create the variable stability responses and produce the control stick feel characteristics.

• Removal of the front-seat flight controls and installation of an electro-hydraulic artificial feel system, i.e., control stick and rudder pedals, in the front cockpit. This was needed to duplicate the simulated aircraft's feel characteristics and to direct the front seat pilot's stick inputs to the computer (the rear seat pilot's controls remained attached directly to the control surfaces as on a normal T-33).

• Installation of hydraulic actuators connected to the elevator, ailerons, and rudder. These are controlled by the computers to make the aircraft respond in the desired manner.

• Installation of variable stability gain controls in the rear cockpit. These controls are similar to the volume controls on a television or radio and are connected directly to the variable stability computer. They allow the safety pilot to adjust the aircraft flight characteristics and control stick's feel. Thus the safety pilot can increase or decrease the abruptness of the aircraft response, make the stick feel light or heavy, and make oscillations die out quickly or slowly.

NT-33 over Edwards Air Force Base on a test pilot instruction flight in the mid-1980s. (USAF Photo)

PART I: IN-FLIGHT SIMULATORS

One of the unique modifications to the NT-33 was the installation of tip-tank mounted speed brakes. (USAF Photo)

The aircraft made its first flight in this modified configuration in February 1957. The flight was completed without a single maintenance write-up, a significant accomplishment considering that the aircraft had not flown for over two years while it was being modified. A year of checkout flights followed in which the operation of the variable stability system was checked and the full capabilities of the aircraft were verified.

The NT-33A suffered its first mishap in July 1958 that could have ended its career. While attempting a landing on a short wet runway with high gusty winds, the aircraft skidded and hit the concrete base of a runway light with one of its main landing gears. The impact resulted in the wing spar being twisted beyond repair. A new wing was obtained and modified as was the original. The aircraft was flying again in October.

The first of many upgrades to the variable stability system was performed early in 1959 when the vacuum tube circuitry was replaced with germanium transistors.

The NT-33A finally performed its first research program in November and December of 1959. It was ferried to the Air Force Flight Test Center at Edwards AFB in California where it performed a generic study of the predicted handling qualities of lifting body re-entry vehicles. In less than six weeks, 100 hours of test flights were performed, a tribute to the rugged design of all the modifications.

During the summer of 1960, the first of many unique program was performed, this one to support development of the X-15 research aircraft (the word "unique" tends to be overused, because many NT-33A research programs incorporate some new capability or use the aircraft in some way not envisioned in the original design). It was desired to simulate the X-15 during descent. The simulation run had to begin when the X-15 was in a ballistic reentry at 0g, supersonic speed, and 70,000 feet altitude! During the evaluation, the pilot had to perform a wings-level pull-up, reaching a maximum of 4g. The simulation was accomplished by tricking the pilot, who flew the simulation under a hood, by reference to aircraft instruments only. The flight profile, which lasted about 100 seconds, was programmed on the NT-33's computers, starting at the initial altitude and Mach number. The NT-33's flight instruments were controlled by the computer. As the simulated X-15 descended, the NT-33's instruments indicated the X-15's condition. The handling qualities changed drastically as the X-15 descended into denser air and slowed. This was accomplished by changing the NT-33's variable stability system gains as a function of time to reflect the X-15's characteristics at any time during the data run. To produce the g's needed to simulate the sustained pull up, the NT-33A slowly banked in a turn, but the pilot's attitude instrument indicated a wings-level pull up! Many of the test pilots who eventually flew the X-15 participated in this simulation.

During 1961 and 1962, the NT-33A underwent another system upgrade. This time, the standard tip tanks were replaced with ones modified into drag brakes. The back end of each tank was rebuilt with hydraulically-actuated surfaces that could be extended into the air stream to produce high drag. This modification would be required for future simulations of lifting body reentry vehicles that flew very steep approaches.

During 1962 and 1963, the ground simulator capability was developed in which the aircraft became part of a ground-based flight simulator. The basic aerodynamics of the NT-33A were programmed on an auxiliary computer that connected on the ground to the NT-33A's computer system. With these two computers working together, the front-seat pilot could now sit in the NT-33A on the ground and fly the simulation that he would later fly in the air. This capability allowed most of the simulation to be checked on the ground, greatly reducing setup time and costs for each experiment. It also allowed the results of the NT-33A simulation to be compared more easily with other ground-based simulations, greatly enhancing the credibility of test results.

ONE-OF-A-KIND RESEARCH AIRCRAFT

A number of generic handling qualities experiments were performed in 1963 and 1964. These were performed to determine acceptable and unacceptable handling characteristics for various sizes of aircraft. In 1965, simulations were again performed to support the development of lifting body vehicles. This time, the M2F2, one of the early predecessors of the X-24, was simulated. In 1966, the X-24 itself was simulated.

By the late 1960s, it was realized that future military aircraft would have to make extensive use of stability augmentation in order to retain acceptable handling qualities while performing their missions over tremendous ranges of airspeed, altitude, weight, external loading, and center of gravity locations. The late 1960s was spent performing many research programs investigating the use of relaxed static stability and electronic stability augmentation. This research contributed directly to the design and development of the F-15, F-16, and YF-17, which were to follow in a few years.

Another new capability was developed during 1971. The controls for the tip-tank-mounted speedbrakes were modified so they could be deployed asymmetrically to produce yaw. By then counteracting the yaw with the rudder, a net side force resulted! In late 1971, the NT-33A participated in the A-9/A-10 flyoff competition by simulating both aircraft. The side force capability was essential for simulating the A-9, because the A-9 used a similar sideforce generating scheme to allow precision aiming for ground attack.

The NT-33A's contribution to current fighter aircraft continued by simulating the F-15 shortly before it made its first flight in 1972. In 1973, an in-flight refueling probe was installed to allow in-flight refueling tasks to be evaluated (in-flight refueling is a very demanding piloting task...if an aircraft has a poor handling characteristic, in-flight refueling is one of the tasks where the problem will probably show up). Also in 1973, a hydraulically actuated side stick was installed to support YF-16 development. Several handling qualities deficiencies in the YF-16 design were predicted and corrections to the flight control system were made. In 1974, a simulation of the YF-17 was performed. On one landing, an "explosively" unstable oscillation began, resulting in an extensive redesign of the YF-17s flight control system before the YF-17 ever made its first flight.

In 1975, both the Air Force and Navy test pilot schools began using the NT-33A to train new test pilots. The in-flight simulator is an ideal tool for this purpose, because it's flight characteristics can be changed easily. It is essential for a test pilot to describe what he sees and feels when testing a new aircraft. Although test pilots train in a wide variety of aircraft, understanding subtle differences can be difficult when going from one aircraft to another. Using an in-flight simulator, these subtle effects can be exaggerated while holding all other characteristics constant, thus helping the new test pilot to identify them and understand their significance. The NT-33A is also used to teach flight control system design. New test pilots see the effects of various types of stability augmentation on the flight characteristics of a baseline aircraft being simulated. They also fly the aircraft with both a centerstick and sidestick to experience the differences. Last, recent problems encountered with real aircraft can be demonstrated, as well as the corrective actions that were taken to correct the problem.

The NT-33A had its closest bout with disaster in 1976 while performing a research project near St. Louis. As the aircraft was climbing to the test area, the wings began to flutter and the tip tanks snapped off, taking with them outboard sections of the wing. Ready to eject, the pilots performed a controllability check and determined the aircraft still had enough control power to attempt a landing. The

NT-33 performing mid-air refueling from a Navy KA-3 tanker in the mid-1970s. (Calspan Photo)

Air Force came very close to recommending the NT-33A be retired following this mishap. Fortunately, farsighted people within both Government and industry saw the ongoing value of the aircraft and convinced the Air Force to rebuild it and keep it flying. The aircraft was disassembled and the wing was replaced. However, to keep the cost down, the tip tanks were not modified into drag brakes because there was no immediate need for them at the time (nor did the need materialize again and the aircraft retained standard tip tanks ever since).

In 1978, the Navy was developing the F-18, which was based on the YF-17. When problems with roll control developed, the NT-33A was used to simulate the F-18 and help develop a correction to the flight control system.

A programmable head-up display (HUD) was installed in 1979. This device projects the attitude and other essential flight information in front of the pilot so he can keep his head out of the cockpit. An evaluation was performed then of several different HUD symbologies. Since this system was installed, most test programs have included the HUD display of the aircraft being simulated.

The NT-33 remained active through the 1980s. Many research programs were flown in attempts to determine why modern aircraft design techniques often still failed to predict poor flying qualities. With the introduction of digital computers as the heart of the flight control system on most new aircraft, new problems emerged that needed to be resolved. Specific aircraft developments that were supported were the AFTI/F-16 (a special test aircraft with many unconventional flight control capabilities), the Lavi (a fighter to be built by Israel, eventually canceled), the JAS-39 Gripen (a fighter built by Sweden, currently entering service), and the YF-22. During 1991 and 1992, the NT-33A performed research programs making extensive use of the programmable HUD. Despite the fact that all current Air Force fighter aircraft use HUDs, there is no standard display and the HUD may not be used as a primary flight instrument while performing instrument flight. Extensive flying was performed to evaluate proposed display configurations to help establish a standard HUD format for all Air Force fighter aircraft.

The NT-33's last big simulation program was an evaluation of the Light Combat Aircraft, a small fighter being designed for the Indian Air Force. This was performed in 1995.

With over 35 years of dependable service, the NT-33A remains active in mid-1995. Although technical support, spare parts, and engines are no longer available through the Air Force depots, adequate stocks of parts have been set aside to keep the aircraft flying and performing much-needed research until it is replaced by the VISTA/NF-16D in-flight simulator.

PART I: IN-FLIGHT SIMULATORS

TU-154M In-Flight Simulator

AIRCRAFT TYPE: Tupolev-154M
MISSION: In-Flight Simulation, Research
OPERATING PERIOD: 1987 - Present
RESPONSIBLE AGENCY: Flight Research Institute, Zhukovsky, Russia

THE STORY

The Russian Tu-154M in-flight simulator grew out of an earlier in-flight simulation effort that supported Russia's Buran space shuttle. In the late 1970s and early 1980s, four in-flight simulators were developed by the Flight Research Institute, located in Zhukovsky, Russia, just outside of Moscow. These aircraft, modified Tu-154B airliners, were used exclusively to simulate the Buran. Their purpose was to develop the flight control laws and control mechanization for the Buran, to allow piloted evaluation of the anticipated flight characteristics, and to train Buran pilots to use the special techniques they would need.

Based on the success of the Tu-154B in-flight simulators and realizing the value of such a specialized aircraft to support other research, the Flight Research Institute built the Tu-154M and has operated it since 1987. The Tu-154M was designed as a general purpose in-flight simulator to support the development of other new aircraft and to perform other flight research such as:

- Developing digital and analog flight control and engine control systems

- Evaluating new pilot hand controllers, including sidesticks, control wheels, and thrust levers

- Developing certification standards and new handling qualities criteria, i.e., determining how a generic class of aircraft should respond to pilot control

- Performing human factors research and pilot workload experiments

- Researching head-up and head-down symbology and format

- Developing special piloting techniques

The Tu-154M utilizes the normal cockpit, with the evaluation pilot sitting in the left seat and the safety pilot in the right. The safety pilot has conventional mechanical controls and instruments. He is warned of any imminent system failures or system limits and can take control of the aircraft by pressing any of several buttons located on his controls or by moving his control wheel past certain thresholds. The only automatic system disengagement is if the angle-of-attack or g-load approach limits.

From the outside, the Tu-154M in-flight simulator appears similar to an Aeroflot airliner. (Flight Research Institute Photo)

ONE-OF-A-KIND RESEARCH AIRCRAFT

Tu-154M inflight simulator cockpit. Note the heads-up display, CRT display, experimental control yoke and side stick. (Flight Research Institute Photo)

The evaluation pilot sits in the left seat and controls the aircraft through completely separate digital or analog fly-by-wire controls. Any of these controls can be changed for particular experiments. A wide variety of feel characteristics can be programmed in advance and varied in flight as needed.

The left-seat displays consist of a head-down and a head-up display. The head-down display is a modified commercial color computer monitor. Instruments can be drawn on the monitor, or new, experimental displays can be depicted. The head-up display was adapted from a fighter aircraft unit, and mounted inverted from the ceiling. The displays for both units are generated using a commercial desk-top computer. Different preprogrammed displays can be selected in flight, and the computer programs that compute the displays can be modified in flight as needed. Other signals, such as the video from a forward looking infrared camera, can be blended with the head-up display.

The aircraft being simulated is modeled on a microVAX-2 digital computer. This includes all computations for the aerodynamics, flight control laws, engine response, and controller feel characteristics. In addition, the Tu-154M features a telemetry system linked to a ground-based computer system. When additional computer power is needed, flight data is transmitted to the ground. Data is then processed and sent back to the aircraft within milliseconds for use in the simulation. This system also allows for engineers on the ground to monitor the aircraft's performance.

Normal crew on the Tu-154M consists of seven people: two pilots, a navigator, a simulation system engineer, a project engineer, and two specialty engineers or observers as needed by the experiment.

The Tu-154M has been used to support the development of the Antonov-70 turboprop transport, and airliners including the wide-body Ilushin-96 and the advanced technology Tupolev-204. It also performed research programs to develop new hand controllers, investigate new displays and flight control laws, and examined the ranges of acceptable handling characteristics.

PART I: IN-FLIGHT SIMULATORS

VFW-614 ATTAS In-Flight Simulator

AIRCRAFT TYPE: Vereinigte Flugtechnische Werke VFW-614
MISSION: In-Flight Simulation and Aerospace Research
OPERATING PERIOD: 1987 - Present
RESPONSIBLE AGENCY: Deutsche Forschungsanstalt für Luft- und Raumfahrt (DLR, German Aerospace Research Establishment)
PRIMARY CONTRACTOR: Messerschmitt-Boelkow-Blohm (MBB)

THE STORY

The ATTAS is an in-flight simulator developed and operated by DLR at their facility at Braunschweig, Germany. ATTAS stands for Advanced Technologies Testing Aircraft System. It was developed in the mid-1980s for the following purposes:

- In-flight simulation of specific aircraft
- Human factors and pilot workload research
- Flying qualities research and control system optimization
- Air traffic control development
- Testing avionics and other hardware components

The ATTAS was built from a Vereinigte Flugtechnische Werke VFW-614 commercial 44-passenger short-haul transport aircraft. Only about twenty 614s were built and production ended in 1978 when VFW merged with Fokker. The aircraft completed were operated by the German Luftwaffe as VIP transports and by airlines in Denmark and France.

ATTAS was built from the 17th aircraft produced, provided to DLR by the Luftwaffe. The VFW-614 was well suited for development into an in-flight simulator. Designed as a jet age DC-3, it was rugged, simple, and economical to operate. It had a cruising speed of 285 knots and cruising altitude of 30,000 feet, yet had a low landing speed of 100 knots. With full fuel, it could still carry about 7000 lbs. of equipment. Despite the small number of VFW-614s built, DLR acquired a large supply of spare parts to keep the ATTAS flying for many years.

ATTAS development started in 1981 when DLR let a contract to MBB in Bremen for most of the major modifications. These included replacing the standard flaps with ones split into three sections that could be operated separately, and installing new actuators, engine sensors, antenna mounts, and electrical power systems. DLR developed the computer system, cockpits, instrumentation, avionics integration, telemetry system, and the 16-foot long carbon fiber nose boom containing sensors for airspeed and angles of attack and sideslip.

The left seat was modified into the evaluation pilot's seat. Standard mechanical instruments were replaced with two cathode ray tube displays. The center control wheel was replaced with a fly-by-wire side stick whose characteristics can be programmed to match those of virtually any hypothetical or actual stick. The safety pilot sits in the right seat. He has controls to disengage the evaluation pilot quickly should a potentially dangerous condition develop. An automatic trim system keeps the aircraft in trim while being flown by the evaluation pilot. This minimizes transients when the safety pilot has to take over control.

In addition to the front evaluation pilot station, another evaluation cockpit is located in the passenger compartment. At this station, futuristic cockpit designs and displays can be tested. Planned uses include development and testing of voice control, touch screens, and direct data links to air traffic control. The evaluation pilot has no outside view when flying from this cockpit.

The computer system consists of seven Loral/Rolm MSE 14 and Hawk 32 digital computers. An optical data bus links the computers to the pilot's controls and displays.

Another essential part of the ATTAS system does not fly, and is not even located in the aircraft. This is a ground-based simulator facility that duplicates the ATTAS's computer system and evaluation cockpit. This facility is used to program and check new simulations on the ground while the aircraft is off flying other programs.

The ATTAS was delivered to DLR with all airframe modifications in September 1985. DLR then installed the remaining systems. The completed ATTAS made its first flight in December 1986.

Research programs and simulations performed to date on ATTAS include simulations of the Airbus A300 airliner, the Hermes re-entry vehicle (this was to be the European space shuttle, but was subsequently canceled) and the Indonesian designed N250 regional transport, comparison of in-flight and ground-based simulations, and improvement of parameter identification techniques (the means by which the values of aircraft aerodynamics parameters are derived from flight test results).

Left: ATTAS in-flight simulator with flaps extended in order to produce direct lift. Right: The ATTAS in-flight simulator was built from a VFW-614 twin-jet transport. (DLR Photos)

ONE-OF-A-KIND RESEARCH AIRCRAFT

Calspan Learjet In-Flight Simulators

AIRCRAFT TYPE: Learjet Model 24, Learjet Model 25
MISSION: Test Pilot Training, Research
OPERATING PERIOD: 1981 - Present
RESPONSIBLE AGENCY: Calspan Corporation

THE STORY

It has often been said that a picture is worth a thousand words. Nowhere is this more true than when teaching test pilots and engineers how to distinguish between the many individual flight characteristics that when taken together, describe how an aircraft flies. The interactions of aircraft dynamics are very complex and difficult to separate and explain, except when demonstrated using an in-flight simulator. The two Calspan Variable Stability Learjets are probably the world's best known test aircraft used for training test pilots and engineers to evaluate flight characteristics.

Over the years, engineers have developed complex analysis techniques to look at flight characteristics in a mathematical sense. These techniques are fine to use on the ground for analyzing and categorizing different effects, but they do not show the pilot what the aircraft will feel like when in the air. They also do not help the pilot differentiate between the many complex interactions of different characteristics. The Calspan Learjets bridge the gap between theory and practice.

The first Calspan Learjet was developed in the late 1970s to replace the two aging B-26 aircraft used at the U.S. Air Force and U.S. Navy test pilot schools. It was recognized by the early 1970s that the B-26s could not be kept in service indefinitely. A study was performed in 1974 to recommend a suitable replacement, but no military aircraft that met all the requirements could be identified. The study was expanded to include civil aircraft with the Learjet being selected.

The primary reasons the Learjet was selected were its structural strength (positive 4.4Gs to negative 1.76Gs) and high thrust-to-weight ratio. Further studies showed that the Learjet could be modified with installation of all needed computer, recording, and hydraulic systems, and placement of an observer seat and additional test engineer stations. With its large normal acceleration, speed, and altitude envelopes, the Learjet could maneuver almost as well as many high-performance fighters that the student test pilots would eventually test.

The Air Force Test Pilot School awarded a contract to Calspan to design and build the aircraft. Under the terms of the contract, Calspan would purchase and own the aircraft and the school would own the unique systems and pay for the modification. Calspan chose to purchase one of the Model 24 test aircraft from Gates Learjet. This particular aircraft was chosen because it already included some of the instrumentation and wiring required, had few flight hours, and contained little interior decoration (most of which would have been discarded, anyway).

The aircraft was completed in 1981, and given the new registration number N101VS. The only external differences that distinguish this aircraft from a standard Learjet are the addition of a second angle-of-attack vane on each side of the nose, a sideslip vane under the nose, and a bulge in the ventral fin to house the rudder actuator. Internally, the right seat was converted into the student seat which has a fly-by-wire center stick, rudder pedals, and a side stick. The center stick and rudder pedals are programmable to produce a wide range of feel characteristics. To save costs, the Air Force Test Pilot School provided the side stick assembly, obtaining it from a convenient location, the NF-104 on static display right outside their door!

A short developmental flight test program requiring 17 flights was conducted over the next ten weeks. The flights were to test engaging and disengaging the variable stability system, perform envelope expansion and flutter clearance flights with the variable stability system engaged and disengaged, determine the range of aerodynamic parameters that could be produced, and finally, develop the actual configurations that would be demonstrated.

The aircraft dynamics that the Learjet is to produce are computed using an analog computer. There are 64 values in the computer that must be set to determine the dynamics for any given configuration. These 64 values, or system gains, are controlled by a digital configuration control system. Up to 128 complete configurations can be stored in the computer. To change between any of the previously programmed configurations, the safety pilot simply selects the configuration number or can step through the configurations in the order they are stored. Compare this to the system in the NT-33A in which over 35 dials must be set manually to define a configuration!

If a student has trouble seeing a subtle effect in any given configuration, the safety pilot can adjust specific system gains to make that effect more pronounced. If a modified configuration has merit

The first Calspan variable-stability Learjet flies over the Navy Test Pilot School. (U.S. Navy Photo)

PART I: IN-FLIGHT SIMULATORS

Calspan Learjet number two is easily distinguished by its additional two windows. (Calspan Photo)

for showing to other students, it can be saved in the digital configuration control system for later retrieval.

N101VS performed its first student instruction session at the Naval Test Pilot School in Spring 1981, quickly followed by deployments to the Air Force Test Pilot School. In a short time, NASA, the FAA, and several aircraft manufacturers began buying similar instruction time for their pilots and engineers. Demand for the aircraft was so great that Calspan built a second Variable Stability Learjet and began operating it in spring 1991.

The second Learjet, a Model 25, was a used corporate aircraft bought on the civilian market. It appropriately was re-registered as N102VS. The Model 25 was selected because it has a longer fuselage to house additional equipment for future growth and greater range that would allow flights to Europe. To save development costs the variable stability system was virtually a clone of the original system.

N102VS soon became a common sight throughout Europe, training pilots and engineers at the Royal Air Force's Empire Test Pilot School and the private International Test Pilot School in England, the French Air Force's test pilot school, and SAAB in Sweden. Its computer capability has been expanded to permit use as an in-flight simulator to support commercial aircraft programs. In this role, it has supported new aircraft developments for Cessna, SAAB of Sweden, and Nusantara Aircraft of Indonesia.

Just exactly how is an in-flight simulator used to train test pilots? How does it put theory into practice? First, the students have a lengthy series of classes on classical and modern stability and control. These are comparable to graduate-level engineering courses. They learn or relearn the analysis techniques used to study and predict aircraft motions (most test pilots have engineering or science degrees, and many already hold Masters degrees). Before each flight, they review with the safety pilot each configuration that they will see on the flight. They review the equations that predict what they will see and look at graphical representations for each configuration. When they fly, they are reminded of these graphical representations prior to sampling each set of dynamics, and reminded how the graphic changes as each different configuration is tried.

Once in the air, the demonstration may start with a nice flying aircraft. The test pilot flies this for a few minutes to get the feel for the flight characteristics. The instructor pilot may then change the short period frequency. This is the quick pitch oscillation that follows a sharp stick input. By making the frequency faster, the aircraft responds faster, possibly to the point that it now feels uncomfortable. Response to turbulence may also be uncomfortable. Next, the damping is increased so that the motion dies out faster. The aircraft again feels comfortable, but responds much quicker. Next, the stick gearing is decreased so that it takes more stick motion to produce the same pitch rate. The aircraft now feels sluggish because more stick is required to obtain the same motion, but the response after releasing the stick is unchanged. Next, the center of gravity is moved forward, and again the aircraft becomes very sluggish. Now, moving the center of gravity aft, the stick becomes very light and pitch damping decreases. The pilot finds it very difficult to point the nose where desired. Eventually the aircraft becomes unstable after the center of gravity passes a certain point. How many pilots get the opportunity to try flying an unstable aircraft? These are but a few of the one hundred or so configurations demonstrated to teach stability, controllability, and handling qualities.

All these instruction flights are done in relative safety, since nothing external is ever changed on the Learjet. The variable stability system makes the standard control surfaces do all the work so that the aircraft responds to the student's commands according to what has been programmed in the computer. The safety pilot is always ready to resume control if the student has any problems.

ONE-OF-A-KIND RESEARCH AIRCRAFT

Jetstar General Purpose Airborne Simulator

AIRCRAFT TYPE: Lockheed JetStar
MISSION: In-Flight Simulation/General Flight Research
OPERATING PERIOD: 1965 - Present
RESPONSIBLE AGENCY: NASA Dryden Flight Research Center
PRIMARY CONTRACTOR: Cornell Aeronautical Laboratory

THE STORY

In the early 1960s, one of NASA's major undertakings was to support development of the supersonic transport. With advances being made in the development of in-flight simulators, NASA decided it needed its own in-flight simulator to support the supersonic transport. Thus, in January 1963, NASA purchased a new Lockheed JetStar executive aircraft for eventual conversion into an in-flight simulator. Experiments planned were to evaluate aircraft flying qualities, develop automatic and manual control systems, optimize instrument displays, and develop training techniques. The JetStar was chosen because it was capable of flying faster than 550 miles per hour at altitudes up to 40,000 feet.

A contract was awarded to Cornell Aeronautical Laboratory (now Calspan Corporation) in Buffalo, New York in June 1964 to design, develop, and install the systems that would convert the JetStar into the General Purpose Airborne Simulator, or GPAS. Equipment consisted of an airborne quality analog computer that used model-following control laws, data recording equipment, and hydraulically-powered flight controls for the left-seat pilot. The right-seat pilot served as the safety pilot. The system could control the pitch, roll, yaw, and forward motions of the aircraft independently of each other. In addition, there was a seat for a test engineer who operated the computer and set each configuration to be flown, and seats for up to two observers. Following modifications, the GPAS was delivered to NASA Dryden in November 1965.

A series of research programs were flown through the late 1960s. These included:

• simulating the XB-70 Valkyrie to evaluate flying qualities and to train pilots who would fly it

• simulating the HL-10 lifting body, a predecessor to the X-24. Unfortunately, this program ended after only two flights because the GPAS was unable to duplicate the rapid motions of the wingless HL-10

• studying the requirements for visual and motion cues for ground-based simulators (Simulators, even ones capable of large motions, cannot completely duplicate all the motions of a real aircraft. Their motion is "washed out" to keep the cockpit within limits of the motion system. How the motion is produced, and how the pilot perceives it has tremendous impact on how the pilot rates the aircraft's flight characteristics. To do this experiment, various motion schemes as used in

GPAS aircraft performing laminar flow studies. Notice that the wing-mounted fuel tanks have been removed. (NASA Photo)

PART I: IN-FLIGHT SIMULATORS

The GPAS with a scale model unducted fan mounted on the fuselage pylon. Also note the instrumentation boom mounted below the nose. (NASA Photo)

ONE-OF-A-KIND RESEARCH AIRCRAFT

The GPAS aircraft, early in its career, prior to the addition of its fuselage-mounted pylon. (NASA Photo)

ground-based simulators were duplicated in the GPAS. The pilot flew using instruments only and restricted outside visibility while the GPAS maneuvered as would the cockpit on a ground-based simulator.)

Starting in the summer of 1971, many upgrades to the GPAS's capabilities were made. These included modifying the flap into a direct lift flap and adding a side force generator, and installing a new flight director system and a new data recording system.

In 1972, the GPAS began a series of test flights to determine acceptable passenger ride qualities (pilots are not the only ones who care how an aircraft responds . . . if passengers don't like it either, they won't want to ride in it). Thirty NASA Dryden employees served as test subjects, performing tasks such as reading, writing, and eating airline meals while the GPAS flew a typical short-haul flight profile. The test subjects rated their abilities to perform these tasks while the aircraft's instrumentation system recorded all aircraft motions. Results were compared with similar assessments made by a number of passengers on scheduled airliners. Following this program, the GPAS performed a series of data gathering flights to determine the aircraft's response to turbulence in order to develop methods to simulate turbulence response accurately. The analog computer then was replaced with a digital computer system.

In 1972, NASA was studying the feasibility of developing a dedicated in-flight simulator for the Space Shuttle. To gain data and have a realistic experience base against which to evaluate proposals, the GPAS was used to simulate the Space Shuttle during final approach. Results from this test aided the eventual development of the Shuttle Training Aircraft.

Soon after the energy crisis struck in 1973, NASA became involved in efforts to make transport aircraft more energy efficient. Managed by NASA's Langley Research Center in Virginia, the Aircraft Energy Efficiency Program had the goal of developing technologies to reduce fuel consumption from 20 to 40 percent by 1985. One portion of this effort was performed on the GPAS: studying the effects of insect impacts. It had long been recognized that the surface roughness caused by such impacts was a source of drag. In fact, the low drag advantage of laminar flow wings then being developed would be lost because the rough surface resulting from bug impacts could ruin the smooth laminar flow.

This study examined different materials and methods that could prevent insect impacts from sticking to the wing leading edge and ruining the smooth laminar flow. Sections of the wing leading edge were coated with four different materials: Teflon tape, Teflon spray, windshield coating, and radome rain repellent. A fifth section of bare aluminum was used as the control section. Instrumentation probes were installed above the panels to measure the airflow and determine if it changed from laminar to turbulent. In another test, a pressurized water system was installed with small nozzles every four inches. This system was evaluated to determine its effectiveness in washing the bugs off in flight. It was determined that the surface could be kept clean as long as it was kept wet.

In 1973 and 1974, several "ride-smoothing" flight control systems were evaluated. This system sensed when the aircraft was bouncing in turbulence and automatically moved the flight controls to maintain the aircraft as level as possible. The system worked by sensing aircraft motion. If there was no corresponding command from the pilot, the computer assumed the motion was the result of turbulence and automatically moved the flight controls to compensate. During one of these flights, the aircraft was seriously damaged when it exceeded the maximum allowable g-force and it was retired. It was eventually decided to replace the wing and return the GPAS to service. However, the variable stability system was removed so it would never again function as an in-flight simulator.

PART I: IN-FLIGHT SIMULATORS

As the Space Shuttle was preparing to begin flight testing and launch operations, the GPAS was used to verify operation of the Microwave Scanning Beam Landing System. This system provides very precise navigation signals to the Shuttle to guide it to the runway. The receiver and display hardware were installed in the GPAS and simulated Shuttle approaches were flown to confirm operation of the system. Tests were performed at Edwards Air Force Base in 1976 prior to the Shuttle being launched from the Boeing 747 carrier aircraft, and then at Kennedy Space Center in 1978 and at White Sands in 1979 to verify operation of those systems.

By the early 1980s, NASA's Aircraft Energy Efficient Program was investigating new propeller designs that promised turboprop efficiencies while flying at near-jet cruise speeds (.8 Mach at 35,000 ft). These props, typically consisting of eight or ten highly swept blades, were called prop fans. Fuel savings of up to 40 percent were predicted. However, excessive noise and vibration problems had to be overcome. Wind tunnels proved unacceptable to gather the needed data. The GPAS was enlisted again, this time as a flight test vehicle for scale models of these prop fans.

A three foot pylon was installed on top of the GPAS' fuselage, atop which was a 195 horsepower turbine engine. A variety of scale model prop fans were attached and tested in flight. Typical prop size was about 2 feet in diameter, roughly one seventh scale models. The aircraft was also equipped with 28 flush-mounted microphones, as well as skin accelerometers to measure vibration. Numerous models were tested over a two year period. The final flight, performed in September 1982, was accompanied by a NASA Learjet in close formation, equipped with microphones to make noise measurements from distances up to 800 feet.

In 1983, the GPAS began another test program supporting laminar flow research. Extensive modifications, requiring about a year, added a suction system to the wing leading edge. This kept the air flowing smoothly over the surface and greatly reduced drag. Different mechanizations were used on each wing. The left wing had 27 spanwise slots, each 0.003 inches wide. The right wing used approximately 1 million holes, each with a diameter of 0.0025 inches. A water/glycol mixture, verified in 1976 tests, was sprayed on the wing to keep it free of insects, ice, soot, and industrial pollution. In May 1984, while performing a system acceptance flight for this program, the GPAS flew its 1000th flight.

Under laboratory conditions, this system demonstrated 25 to 40 percent fuel savings. The real test would be a demonstration under typical airline operating conditions. To perform this, the GPAS was operated in and out of various major airports around the country during the summer of 1985. Two to four flights were performed each day, around mid-day and dusk, which were peak insect activity times. Because these flights were intended to replicate a typical airline operation, major airlines served as consultants to insure the test conditions were as realistic as possible.

The GPAS remains an active research aircraft in the mid-1990s, still operating out of NASA's Dryden Flight Research Center after thirty years.

PART II
TESTBEDS

The most recognizable (and by far the most numerous) one-of-a-kind aircraft are the so-called testbed aircraft. In this particular case, an existing production, or pre-production, aircraft is modified with a new technology to be tested or verified.

The modification can be manifested in many ways. First, there's the airframe itself which can be modified for ordnance or engine testing, or the wings can be changed to test new aerodynamic concepts. The fuselage can be altered with the addition of new blisters, sensors, antennae, etc. Also, these aircraft can be utilized to test new guidance and flight regimes with changes to internal software. And so the list of modifications goes, it's practically endless.

It's almost impossible to discuss, or even find, all the testbed aircraft through the years because the practice has been going on since the first aircraft was built. And it continues in the 1990s. One of the most recent examples is the use of the new YF-23 fighter prototype for testbed purposes.

The obvious advantages of the testbed concept are economic, schedule, and safety. Modifying an existing aircraft, where the test function can be accomplished by merely making a modification, really gets the job done for minimum dollars in a short time, with least risk. Certainly, there are huge advantages in these days of decreasing military spending and tight civilian budgets.

PART II: TESTBEDS

A-5A Vigilante SST Simulation Testbed

AIRCRAFT TYPE: A-5A Vigilante
MISSION: Supersonic transport approach studies
OPERATING PERIOD: 1960s
RESPONSIBLE AGENCY: NASA Dryden Flight Research Center

THE STORY

The aerodynamic configuration of the A-5A Vigilante had the perfect shape for the mission that NASA Dryden had planned for it in 1963. The low-aspect-ratio swept-back wing had no ailerons, while the plane's blown flaps were used for low speeds, with the spoilers and rolling tail for high speed simulations. The aircraft also had variable-geometry intakes. The mission of the testbed was to support development of the Supersonic Transport (SST). This particular aircraft was borrowed from the U.S. Navy.

The actual mid-1960s testing was to determine the problems inherent in operating an SST in a dense air traffic network. This was first explored by the subject testbed, with light traffic. Supersonic approaches were eventually flown into the terminal approach and departure zones at Los Angeles International Airport.

The piloting problems associated with flying the supersonic profile appeared to be minor, limited primarily to speed and altitude fluctuations during the high speed, high altitude parts of the profile. There was also a tendency to overshoot during level-off maneuvers from steep portions of the climb. Evaluations using this testbed aircraft revealed only minor problems in merging SST flight operations with normal commercial passenger jets.

This Navy A-5A Vigilante was used by NASA to simulate approaches for the ill-fated American Supersonic Transport program. (NASA Photo)

ONE-OF-A-KIND RESEARCH AIRCRAFT

A-6A Circulation Control Wing (CCW) Testbed

AIRCRAFT TYPE: A-6A
MISSION: Development of low-speed approach capabilities
OPERATING PERIOD: Late 1970s
RESPONSIBLE AGENCY: U.S. Navy
CONTRACTOR: Grumman Aerospace Corporation

THE STORY

A majority of the subsonic wing innovations through the years were directed at two main goals: short take-off and landing performance, and fuel efficiency. The late-1970s U.S. Navy-sponsored Circulation Control Wing (CCW) program definitely concentrated on that first objective with outstanding results.

The program, which was completed in 1979, had a total of 21 test flights which demonstrated a significantly reduced approach speed and reduced landing rollouts. For example, in the final flight test, the A-6A testbed aircraft had a landing roll of only 1,075 feet, whereas the normal landing roll for the aircraft (in the unmodified configuration) is about 1,700 feet. Also demonstrated was an amazingly-slow approach speed of only 78 knots as compared to the standard A-6A speed of 120 knots.

The program management was headed by the David W. Taylor Naval Ship Research and Development Center in Bethesda, Maryland with Grumman Aerospace Corporation being the prime contractor.

Obviously, considerable modifications to the testbed aircraft were necessary in order to accomplishment the large increases in these capabilities, and the changes were very interesting to say the least.

The CCW concept involved the enhancement of the aircraft's lift capabilities through use of the so-called "Coanda" effect on the wing leading edges. Bleed air was tapped from the aircraft's two wingroot-mounted engines and directed through a manifold. From there, the air entered titanium and steel structures on the trailing edge of each wing.

The structures actually gave the trailing edges of the wings a circular cross section. The bleed air entered through a narrow slot along the top of each structure where a sheet was formed which flowed around the surface of the structure. Adhering to the round structure, the bleed air then separated in a downward direction off the trailing edge.

The overall purpose of the complicated arrangement, which increased the overall airflow, or circulation, of the wing, ended up yielding considerable more lift. One advantage of the system came from its capabilities during an engine-out situation. It was found that the CCW concept allowed an evenly-distributed decrease in lift.

When the CCW system was operating in the trailing edge blowing mode, only about 12 percent of each engine's bleed air was tapped. That equated to a maximum flow rate through each nozzle of about 12 pounds per second. The bleed air came from each engine at a temperature of about 700 degrees, and cooled only about 40 degrees before reaching the trailing edge structures.

Since the A-6A was obviously not built from scratch for the CCW mission, it was also necessary to modify other parts of the structure to make the plane totally compatible. First, the horizontal

This Navy A-6A bomber was modified in the late 1970s time period to simulate low-speed approaches. The modified trailing edges are not visible, but the stall fences are. (U.S. Navy Photo)

PART II: TESTBEDS

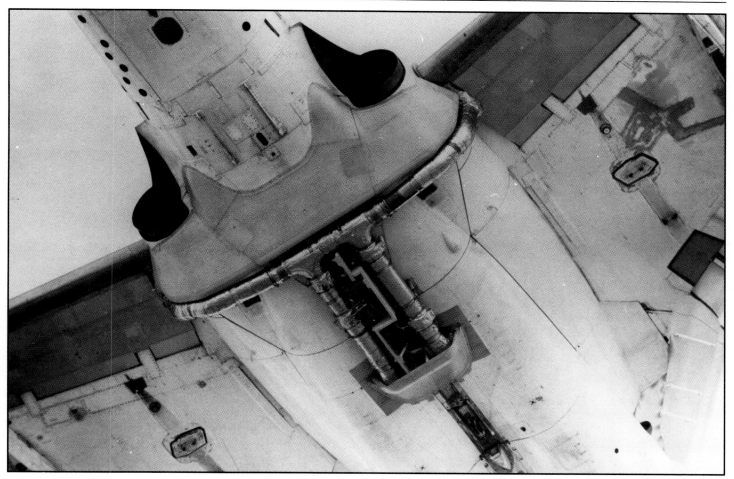

The modifications to this testbed aircraft can be seen under the twin engine exhausts. The concept was to trap bleed air from the exhausts and then direct the air through special slots on the trailing edge of each wing. (U.S. Navy Photo)

stabilizers were reconfigured with increased area and inverted camber, while the wing slats were increased in radius and fixed at 25 degrees. The glove strakes were replaced with fixed Krueger leading edge slats, while enlarged fences were added to each wing outboard of the trailing edge nozzles.

The Navy had long been interested in the CCW technology beginning basic research on the concept in the late 1960s. Then, during the 1970s it was decided to test the concept with a fleet aircraft. The aerodynamics of the A-6, along with the fact that it had twin engines, was instrumental in its selection. The actual contract was awarded to Grumman in 1977 with the final flights occurring in 1979.

The use of this particular production aircraft testbed was obviously selected for application in an aircraft carrier environment. But even with the promising results the program demonstrated, the technology would effectively die with the program.

Unfortunately, the promising CCW concept never appeared in any future aircraft.

ONE-OF-A-KIND RESEARCH AIRCRAFT

B-47 Fly-By-Wire Testbed

AIRCRAFT TYPE: Boeing B-47E
MISSION: Fly-By-Wire Research/Demonstration
OPERATING PERIOD: 1967-1969
RESPONSIBLE AGENCY: Flight Dynamics Laboratory
PRIMARY CONTRACTOR: Hydraulic Research and Manufacturing Company

THE STORY

In the 1950s and 1960s, the design of aircraft was moving toward higher performance, longer range, and more varied mission requirements. To help accomplish this goal, aircraft components and systems had to become lighter, more reliable and maintainable, and easier to update as mission requirements changed.

Aircraft flight controls had long been recognized as one area where new technologies could help meet this goal. The complex system of cables, pushrods, and bellcranks running from the control wheel to the control surfaces was a mechanical nightmare. They required long, straight runs through the fuselage and wings, needed frequent adjustment, were vulnerable to combat damage, and were difficult to make redundant.

Most high performance aircraft already had hydraulic boost on the control surfaces to reduce the high forces the pilot otherwise would have to apply to make the aircraft move. The concept of fly-by-wire was to replace many of the mechanical components that ran from the control wheel or stick to the control surface with an electrical wire. As the pilot moved his controls, electrical signals would be generated and sent to electronically controlled valves at each control surface actuator. The valves would then be opened or closed to make the control surfaces move.

The first in-flight demonstration of a completely fly-by-wire flight control system was performed on a B-47E, serial number 53-2280. This aircraft was operated by the 4950th Test Wing, but little early history of this aircraft could be obtained. It spent its entire career at Wright-Patterson Air Force Base, but it is unknown if it served with an operational bomber wing at the base or with the test wing for its entire career.

Despite the trend to paint distinctive program markings on test aircraft, virtually nothing distinguished 53-2280 from a standard B-47. The only external differences were three small air intakes on the right side of the fuselage just under the horizontal tail (for extra hydraulic cooling), a small window on the front of the nose (behind which a motion picture camera was mounted), and Air Force Systems Command and Aeronautical Systems Division emblems under the canopy.

The purpose of the B-47 Fly-By-Wire program was to develop and demonstrate a fly-by-wire system in the pitch axis to confirm the advantages that had long been surmised. The three phases of this program were:

Phase I - A position sensor was mounted on the normal control column. This sensor produced an electrical signal proportional to the pilot's pitch command. Electrical cables were connected to an electric valve on an elevator actuator installed in parallel to the standard actuator. The first flight occurred in December 1967. Over 45 flight hours were performed, including touch-and-go landings, through August 1968.

Phase II - A side stick controller was installed in place of the control wheel. In addition, aircraft motions such as pitch rate and pitch acceleration were measured and electrical signals pro-

One of the earliest testbed aircraft to investigate fly-by-wire capabilities was this B-47. Presently, this plane is displayed in the Air Force Museum where it has been repainted in operational colors. (USAF Photo)

PART II: TESTBEDS

At the completion of this program the aircraft appeared as shown here. Note the glass window in the nose and the air inlets under the tail still on the aircraft today. (USAF Photo)

portional to their values were added to the pilot's stick input to alter the flight characteristics. The sidestick sensitivity could be adjusted in flight by the flight test engineer to optimize the feel characteristics. These two capabilities made it easier for the pilot to control the aircraft with greater precision. When using the fly-by-wire system, the flying qualities were consistently rated as superior to the standard B-47. A total of 34 hours of evaluations were performed, including tracking ground targets and instrument approaches, between January and May of 1969.

Phase III - A quadruple-redundant electrohydraulic actuator, representative of what might be installed for operational use, replaced the non-redundant system used in the earlier phases. This actuator monitored its own operation, and could detect and compensate for failures. Also, a new side stick having lower force gradients replaced the one used during Phase II. Twelve flight tests were flown for 18 hours of evaluations. During the testing, three different types of failures were intentionally inserted, and the system successfully detected and compensated for all of them.

The B-47 Fly-By-Wire program ended with the last flight in November 1969, when the aircraft was flown to the Air Force Museum for static display. Most of the program equipment was removed, and the aircraft was eventually repainted to represent an operational B-47. Externally, 53-2280 looks like a standard B-47E, except it still retains the nose window and the extra air intakes. Although documentation could not be located, this aircraft certainly participated in other test programs. This is evident by a myriad of unique, unidentified items still installed throughout the interior.

ONE-OF-A-KIND RESEARCH AIRCRAFT

YA-7D DIGITAC Testbed

AIRCRAFT TYPE: YA-7D Corsair II
MISSION: Digital Flight Control Research
OPERATING PERIOD: 1975-1991
RESPONSIBLE AGENCY: Wright Laboratory, Wright-Patterson Air Force Base, Ohio
PRIMARY CONTRACTOR: Honeywell

THE STORY

The DIGITAC A-7 was a test program that demonstrated and helped prove the feasibility of digital flight control. DIGITAC stands for Digital Flight Control for Tactical Fighter Aircraft. In the 1960s, digital computers were becoming smaller, lighter, cheaper, and more reliable. Their potential for performing aircraft flight control computations was obvious. The flight control computer is a critical part of the system that stabilizes high performance aircraft, making them more controllable over a larger flight envelope. Analog computers used in most fighter aircraft were reliable, but were heavy and required a lot of electrical power. As computations grew more complex, so did the amount of circuitry required. Also, system upgrades required new circuit cards to be produced and installed in the computer. Digital computers had the advantage that system upgrades could be performed simply by replacing software.

The Flight Dynamics Laboratory at Wright-Patterson Air Force Base in Ohio was an early advocate for the use of digital flight control. It sponsored several studies that concluded that digital computers could indeed be used and that limitations could be overcome. By 1973, the lab was ready to demonstrate the concept in an actual aircraft. Several potential aircraft were studied, and the A-7 was selected as being best suited for the task.

A YA-7D aircraft was obtained from Edwards Air Force Base. This particular aircraft, 67-14583, was a prototype of the Air Force A-7D version of the Navy's Corsair II. It had last been used at Edwards to test runway barrier systems.

Modifications to the aircraft were made by the Honeywell Avionics Systems Division and included replacing the analog flight control computer with a two-channel digital system. Many of the internal parameters in the digital computers, called system gains, could be adjusted by means of switches and dials in the cockpit. This allowed the pilot to make the aircraft respond differently to identical stick inputs. A built-in test function was developed to allow each computer to monitor itself and the other computer, and to take corrective measures should a problem be detected. These measures included taking itself or the other computer off line if needed. A recording system could record pilot inputs, control surface and aircraft motions and various parameters internal to the computer. The recorder and other needed electronics equipment were installed in a pod under the right wing. For balance, a dummy pod was carried under the left wing.

The DIGITAC made its first flight in February 1975, followed by flights to check all systems and expand the envelope. The aircraft then began its research flying at Edwards Air Force Base. The Flight Test Center flew the aircraft for the Flight Dynamics Lab and Honeywell was retained to maintain the special one-of-a-kind systems installed. The first set of tests, eventually dubbed DIGITAC I, consisted of 92 flight hours. They investigated the necessary computer cycle time (how many times per second the computer had to perform the complete set of computations) and the ability of the built-in test function to detect and respond to partial system failures. This effort also demonstrated the potential for "multimode control laws", in which the flight characteristics are varied for each different piloting task (i.e., making the aircraft fly differently for air-to-air or air-to-ground, than it does during cruise) to decrease the pilots workload and increase his accuracy.

The DIGITAC worked so well and proved to be such a versatile tool that the Air Force Test Pilot school asked to use it as part of their flying curriculum. Thus, in 1976, the school began DIGITAC flights so that new test pilots could experience first-hand the benefits and problems associated with digital flight control. The Flight

The DIGITAC Program used the Corsair II fighter as the testbed aircraft. Besides the high-visibility marking and the test pod being carried, the A-7 appears to be quite normal. (USAF Photo)

PART II: TESTBEDS

During this phase of the DIGITAC Program, the A-7 testbed aircraft had an interesting test pattern on the side of its fuselage. (USAF Photo)

Dynamics Lab continued to fund small research efforts that were performed as student "thesis" projects, a cost-effective way to continue researching this new technology.

Beginning in 1979, the aircraft was upgraded to perform further digital research. This effort, appropriately named DIGITAC II, looked at connecting all avionics, flight control, and sensor components with a digital multiplexing system. The upgrades were performed by Vought Aircraft. A multiplex system connected all components with a cable, called a data bus, replacing hundreds of dedicated wires running throughout the aircraft. Each signal had a certain "time slot" in which it occupies the bus and could be read by every other part of the system. This capability offered tremendous weight savings because of the large number of wires usually needed to route the hundreds of electrical signals normally travelling through an aircraft. It also could decrease the chances of maintenance errors and simplified the routing of cables. Two types of data busses were tested, one using pairs of shielded twisted wire, and one using fiber optic cables. The last part of the upgrade was to allow a movable gunsight to be drawn on the head-up display. This was to demonstrate that the moving sight could make it easier for the pilot to line up on a target, as compared with a static sight. DIGITAC II testing continued through 1981, after which the Air Force Test Pilot School again took over the aircraft.

The DIGITAC III program started in 1988. New computers were installed, and all programming was performed in Ada, the new standard digital computer language adopted by the Department of Defense. This system was evaluated as a potential production upgrade for the A-7 fleet. However, by the time the effort was well under way, the Air Force was planning to retire all A-7s, and the fleet upgrade was canceled. The DIGITAC returned to the Test Pilot School, again with its new systems to demonstrate further these new capabilities.

The Flight Dynamics Lab, which by now became part of the Wright Laboratory, was beginning to plan a DIGITAC IV program in 1991 to investigate further additional digital technologies. However, with the Air Force retiring the A-7 fleet, the Flight Test Center determined that it would be impractical to continue operating a single aircraft without adequate logistic support. The DIGITAC made its last Test Pilot School flight in July 1991, and was then retired. It was in storage in 1994 awaiting eventual static display at Edwards Air Force Base.

ONE-OF-A-KIND RESEARCH AIRCRAFT

B-52 LAMS/CCV Program Testbed

AIRCRAFT TYPE: B-52E
MISSION: Advanced flight control system development
OPERATING PERIOD: LAMS: 1966-1968
CCV: 1971-1974
RESPONSIBLE AGENCIES: Air Force Flight Dynamics Laboratory
CONTRACTOR: Boeing Company

THE STORY (LAMS)

The Load Alleviation and Mode Stabilization (LAMS) Program, carried out by the Air Force Flight Dynamics Laboratory in the mid-1960s, was the first of two programs carried out on this B-52 testbed aircraft. This program was the first of many which would investigate Active Control Technology (ACT) with a number of different aircraft. All the testbed aircraft used were modified production aircraft, and will all be addressed by this book.

The purpose of the LAMS program, though, was to demonstrate the capabilities of an advanced flight control system to alleviate gust loads and control structural modes on a large bomber-type aircraft. The theory behind the program was that by producing structural loads caused by gusts and turbulence, the structure could be made lighter, while fatigue problems could be greatly reduced.

All available control surfaces were used during the LAMS program, with a number of gyros installed along the length of the fuselage, which provided the structural mode rate signals to the flight control system. The two outboard spoiler panels were operated symmetrically around a 15-degree biased position, while the ailerons were used both symmetrically and asymmetrically, while the elevators were used in the normal manner.

To permit ease of development and integration of the new technologies, the pilot's side of the control system was configured as the complete fly-by wire system and operated in parallel with the load alleviation and the basic mechanical systems.

The LAMS system proved to be very successful and reduced the basic airframe wing fatigue damage by about 50 percent. LAMS also demonstrated large decreases in fatigue damage rates at the mid-fuselage stations.

A Direct Lift Control (DLC) study as a part of the LAMS system conducted to demonstrate the desirability of uncoupling the rotational and translational degrees of aircraft motion. Spoilers and symmetrical ailerons were coupled with elevators to implement the DLC program. Flight test results showed that uncoupling pitch and

The testbed B-52 is shown here in its CCV configuration. It was characterized by its large forward probe and two sets of fuselage-mounted canards. Note the high-visibility paint scheme. (USAF Photo)

PART II: TESTBEDS

The CCV NB-52E refueling from a KC-135 tanker. Fly-by-wire controls made the aircraft more stable in turbulence, making in-flight refueling much easier. (USAF Photo)

roll through DLC greatly simplified precise maneuvering required during mid-air refueling and instrument approaches.

Results of the LAMS Program were incorporated into production B-52s to lengthen the life of those aircraft. Much of the results acquired through the LAMS program were also fed into the follow-on CCV program which would employ the same B-52 testbed aircraft.

THE STORY (CCV)

The Control Configured Vehicle (CCV) Program was an outgrowth of the LAMS Program to determine ways and means to integrate new flight control concepts to improve both the aerodynamic and structural design of the vehicle. One of the significant additional results in the program was the fact that the aircraft weight was significantly reduced.

Using the same B-52E LAMS aircraft, the CCV program was initiated in 1971, three years following the conclusion of the LAMS program, and completed in 1974. The term CCV is used to describe a design philosophy in which modern control technology is allowed to impact vehicle design from early in its design cycle.

A number of configuration changes were made to the trusty old testbed aircraft. Two sets of canards were added to the nose of the aircraft, one set being completely horizontal on the side of the body, with the other mounted at about a 45 degree angle under and slightly forward of the horizontal pair. Also added were out-board ailerons and inboard flaperons along with a 2000-pound ballast carried in each of the outboard tip tanks.

The modifications provided for a number of performance improvements and weight reductions. The program aptly demonstrated that the speed of future large aircraft need not be limited in order to

The CCV B-52 sits ingloriously at Davis-Montham AFB, awaiting possible future work in 1988. (Steve Markman Photo)

49

avoid flutter or structural bending. The CCV modifications actually allowed the test aircraft to surpass its design flutter speed. This capability was the first time a manned aircraft was able to make such an accomplishment, i.e., depending on a control system to avoid structure divergence.

The flutter suppression system on the CCV testbed worked by controlled both wing torsion and bending modes, thus providing a degree of redundancy. The flaps did a very good job of controlling the bending mode. The ailerons were also effective in controlling the torsion mod (because of the long, slender wing). With both systems working, the system worked very well. If either system failed, flutter was still prevented because one of the modes was still controlled. This gave the pilot time to get back to a safe speed.

Final Mission
It appeared that the B-52 LAMS/CCV aircraft would languish at Davis-Montham in a well-deserved retirement. Well, for a time it sat there, its red paint scheme slowly fading in the blazing sun.

But then it was asked to perform its final mission, this time it would be the destruction of the venerable Stratofortress.

The mission was a part of an FAA commercial aircraft hardening study designed to make aircraft more survivable to bombs and structural fatigue. The old testbed was loaded with explosives and the TNT was detonated. The tests were accomplished in the spring of 1994. The old Stratofortress gave its life for the cause. Later, what was left would be cut up for scrap, certainly a very unglorious end to an illustrious career.

It served well.

The results of 1994 vulnerability testing to study how an internal explosion damages the structure of a large aircraft. (USAF Photo)

PART II: TESTBEDS

Carrier Testbeds

TYPE OF AIRCRAFT: Boeing B-52, Lockheed L-1011, Boeing 747, Boeing 720, Boeing 707, and Bell 204 Helicopter
MISSIONS: Carriers of various powerplants and payloads, also as serving as test vehicles
TIME PERIOD: 1970s to 1990s
RESPONSIBLE AGENCY: NASA Dryden Flight Research Center/GE
CONTRACTOR: Boeing Company

THE STORIES

NASA B-52 Payload Carrier Aircraft
For years, many manned and unmanned spacecraft have received a free ride to launching altitude from the NASA NB-52A Mother Ship which first flew in 1953. The SN 008 NB-52 has launched the X-24, HL-10, and the X-15 among others. For a number of years, there were two of these aircraft, but the other B-52A (number 003) was retired in 1969 leaving this vintage Stratofortress alone to carry on decade after decade.

This obviously is the oldest B-52 left in flying status with recent activities being the testing of the F-111 crew capsule parachute recovery system, along with also being the launch platform for the commercially developed-Pegasus space booster system. In another recent space application mission, 008 was used for a series of eight tests of a drag chute deployment system which was installed on the space shuttle orbiter.

The aircraft also carried aloft a three-eighths scale model of the F-15 fighter for spin testing. The model was air-launched from the NB-52 at 45,000 feet with commands being transmitted to the unmanned vehicle from the ground. This reduced the amount of high risk spin testing that needed to be performed on actual F-15s.

Other activities have included being the air-launch platform for several remotely piloted vehicles studying aircraft spin-stall characteristics, high angle of attack, and fighter technologies.

The venerable Stratofortress was initially modified with a pylon located between the fuselage and inboard engine under the right wing. All the air-launched vehicles have been were carried on that pylon. In all, 008 has carried out over 900 test missions (as of the early 1990s), with no end in sight.

Although this B-52 has received the majority of the publicity for air-launching various types of test vehicles, there are several other aircraft performing the same mission.

The NASA NB-52 served as the launching aircraft for many space and aeronautical vehicles. Here, it is seen launching the X-15, certainly one of the most-often seen photos of this aircraft. (NASA Photo)

The X-24 lifting body was one of many vehicles carried aloft by the NB-52 (0008) launching aircraft. (NASA Photo)

The NB-52 (0003) carrier aircraft is shown carrying an early M2-F2 lifting body test vehicle. This NB-52 is currently on display at Pima County Air Museum, Tucson, Arizona. (NASA Photo)

PART II: TESTBEDS

The special right side pylon located between the fuselage and inboard engine was the mounting location for the many test vehicles carried by the NB-52 (008). (NASA Photo)

This particular carrier B-52 (003) was the first of the two testbeds, but was retired in 1969 leaving only one to carry on. Here is shown carrying a later version of the X-15. (NASA Photo)

ONE-OF-A-KIND RESEARCH AIRCRAFT

The commercially-developed Pegasus space launch vehicle races away from its NASA NB-52 launch aircraft. (USAF Photo)

This vintage B-52 was modified into a testbed configuration to test the huge TF-39 powerplant for the C-5A Galaxy. The engine was carried on the right inboard pylon. (General Electric Photo)

An almost identical B-52 testbed aircraft was configured to test the engine for the Boeing 747. The JT9D engine was carried in the same inboard rightside position. (Pratt & Whitney Photo)

PART II: TESTBEDS

This Boeing 720 transport is used by Pratt & Whitney Canada to carry a number of different propulsion systems. (Pratt & Whitney Canada)

Lockheed L-1011 Engine Carrier
Another carrier testbed is a Lockheed L-1011 modified airliner, the property of Orbital Sciences Corporation, which took over the launching duties of the Pegasus Launch Vehicle in the mid-1990s. Unlike the NB-52 pylon arrangement, this mother ship carried the Pegasus payload directly under the fuselage.

B-52 Engine Carriers
Two other testbed B-52s have been used for the testing of two new large jet engines. The TF-39 (engine for the Lockheed C-5) and the JT9D (engine for the Boeing 747) were both flight tested on modified early-model B-52s. The installations were basically identical with the huge powerplants mounted on the right inboard engine location. It must be reiterated that these were two different aircraft used on these programs.

Boeing 747 Engine Carrier
But again, the B-52 is not the only engine carrier either with a giant Boeing 747-100 also serving as an engine carrier. The 747 was used to test the General Electric GE90 powerplant which is the prime powerplant for the Boeing 777 transport.

Boeing 720 (Canadian) Engine Carrier
Also north of the border, Pratt & Whitney Canada has used its Boeing 720 flying testbed to test a number of new engine systems including the PW500 engine.

Boeing 707 Engine Carrier
CFM International used a Boeing 707, staged out of Mojave, California, to test new powerplants. During the 1990s, the aircraft tested the CFM56-5B engine which was being considered for various Airbus configurations.

Either a turboprop or pure jet engine can be mounted on the forward fuselage of this testbed transport. (Pratt & Whitney Canada)

ONE-OF-A-KIND RESEARCH AIRCRAFT

Bell 204 Helicopter Model Carrier
You wouldn't expect a helicopter to preform in a testbed capacity, and you can believe that there aren't many of them, but this NASA Dryden helicopter is one notable exception. This versatile testbed started its fourth decade in the mid-1990s.

The main purpose of the testbed through the years has been to serve as a drop platform for a number of free-flight models including the F-4, F-14, F-15, B-1, F-18, F-16XL, and X-29.

Mid-1990s activity included the drop testing a free-flight model of the X-31 research aircraft. The testing evaluated high angle-of-attack flight characteristics including controllability, stall, departure, spin, tumble, and recovery.

Other realms of test activity have included smoke sampling test research, aerial photography, and tail-boom strake aerodynamic research.

Above: This Bell 204 NASA helicopter is used as a carrier testbed aircraft, shown here carrying a model of the X-29 research aircraft. Below: Another aircraft model, this time an F-5, is carried by the Bell 204 helicopter. (NASA Photos)

PART II: TESTBEDS

XC-8A Air Cushion Land System Testbed

AIRCRAFT TYPE: XC-8A
MISSION: Testbed aircraft to demonstrate air cushion landing system (ACLS)
OPERATING PERIOD: 1974 to 1977
RESPONSIBLE AGENCY/S: Air Force Flight Dynamics Laboratory and Canadian Dept. of Industry, Trade and Commerce
CONTRACTOR: Bell

THE STORY

It was a superb example of international cooperation between the United States and Canada for the development of a new aero-space innovation. Such was the mid-1970s test program directed toward the development of an Air Cushion Landing System (ACLS).

The program involved the use of an XC-8A aircraft which was fitted with a huge, air-filled rubber doughnut to serve as a landing gear. The engineering concept was actually quite simple with the donut-shaped ring under the aircraft's fuselage sealing low air pressure under the plane which allowed landings on all types of terrain, as well as even water landings.

The Air Force Flight Dynamics Laboratory at Wright Patterson Air Force Base in Dayton, Ohio and the Canadian Department of Industry, Trade and Commerce jointly sponsored the ACLS program. During the 40-month ACLS test program, the C-8A aircraft logged a total of 84 flights and 144 hours of test time. The aircraft made 34 air cushion take-offs and 39 landings on grass, snow, and hard surfaces, and completed 25 miles of taxi tests on grass, snow, hard and soft surfaces, and even over obstacles.

The XC-8A was delivered by Canada to Wright Patterson in early 1974. Its first air cushion take-offs and landings took place in March and April of 1975. The 4950th Test Wing at Wright Patterson was the conductor of the testing. The cold weather and snow tests were conducted at the Canadian Forces Base Cold Lake, Alberta, and Yellowknife, North West Territories, Canada.

ACLS Program Manager Wallace Buzzard of the Flight Dynamics Lab explained at the time that the flight test proved that an air cushion could be the landing system for future large transport aircraft: "We learned that the ACLS could operate on difficult surfaces – including grass and dirt. The test aircraft also taxied smoothly over a 30-foot long ramp with a nine-inch drop-off, craters six feet in diameter and ditches two feet wide. An aircraft with a conventional landing gear simply can't do those things. It would be seriously damaged, if not wrecked. These are the truly revolutionary feats of the ACLS."

The test program also demonstrated many in-flight inflations and deflations of the air cushion system.

During the taxi tests, test pilots discovered that they could control the aircraft even with cross winds as high as 20 miles per hour. Although the aircraft taxied almost sideways, the pilots learned how to let the wind work for them in moving the aircraft to a particular location.

During all ACLS flight testing, the C-8A weighed 33,000 pounds and carried no cargo. Test engineers indicated, though, that the same tests could have been accomplished with a fully loaded

In a joint U.S./Canadian test program, this XC-8A was modified with an air cushion system for landings on all types of terrain. Note the wingtip floats for water tests. (USAF Photo)

The XC-8A testbed aircraft during maintenance at Wright Patterson AFB. The air cushion landing system and the auxiliary turbine under the right wing root are clearly visible. (USAF Photo)

configuration. Buzzard indicated that an ACLS system could be used to handle aircraft as heavy as three million pounds, along with being competitive with conventional landing systems. He also indicated that the ACLS would be very competitive in cost with conventional systems.

It should be noted that the XC-8A program wasn't the first to test the concept. Earlier, Bell Aerospace Division in Buffalo, New York conducted a similar test program using a CC-115 transport. The system also used a donut-shaped ring mounted underneath. Original flight tests of the system were carried out in the late 1960s using an LA-4 amphibian. The CC-115 test aircraft, so equipped, began its flight tests in 1973.

Although the concept showed considerable promise for a number of applications, the air-cushion concept would never be incorporated into an operational aircraft.

PART II: TESTBEDS

CONVAIR 990 Landing System Research Aircraft

AIRCRAFT TYPE: Convair 990
MISSION: Landing Gear System Development and Testing
OPERATING PERIOD: 1993 - Present
RESPONSIBLE AGENCY: NASA Dryden Flight Research Center

THE STORY

The landing gear on an aircraft is seldom thought of as being on the cutting edge of technology. But when the capabilities of the landing gear reaches its limit, the landing performance of the aircraft also becomes limited. As the operating envelope of the Space Shuttle Orbiter was expanding, its ability to carry heavier loads was limited by touchdown weight and the side loads caused by cross winds. This restricted the Shuttle's ability to retrieve heavy satellites from orbit and limited launch options when carrying heavier loads into space.

To test a new Shuttle landing gear system, NASA developed the Landing System Research Aircraft, or LSRA. It is an extensively modified Convair 990 airliner and carries the designation NASA 810. The LSRA development was the biggest aircraft ever modified at NASA Dryden, as well as the biggest in-house project ever accomplished there. Although new tire designs are routinely tested on aircraft and entire landing gears for new aircraft have even been tested on existing aircraft, the LSRA was the first test aircraft dedicated exclusively to developing and testing landing gears.

The concept for LSRA dates to about 1985. Engineering studies were performed to verify that the concept was viable and that the Convair 990 was strong enough, could accept the needed modifications, and could be landed safely at typical Shuttle speeds of 230-250 mph. Approval for the project was given by the Johnson Space Center in January 1989.

The Convair 990 airframe was obtained from NASA Ames Research Center, where it was used previously to support a variety of test programs. Prior to that it was operated by American Airlines as a passenger transport. The aircraft entered the modification hangar in May 1990 and did not emerge for nearly three years.

Major modifications included the following:

• Installing a truss assembly between the 990's main landing gear to which the test landing gear was mounted (this truss assembly itself was the largest single item ever manufactured in-house at Dryden).

• Removing a section of the main keel and replacing it with two auxiliary keels to maintain structural integrity and to carry the loads from the test landing gear (the truss attached to these two auxiliary keels).

• Installing a computer controlled actuator system that could apply up to 250,000 pounds of force to the test landing gear to drive it downward onto the runway at precise speeds and loads.

• Installing 48 nitrogen bottles and 16 hydraulic accumulators in the aft cargo bay to power the actuators.

• Installing four test consoles inside the cabin: two of which were for instrumentation racks, one for video recording and monitoring, and one for the test director.

The LSRA Convair 990 taxis from the NASA Dryden facility. Note the taxiway behind the vertical tail and the rocket engine test rigs located on the mountain. (NASA Photo)

ONE-OF-A-KIND RESEARCH AIRCRAFT

Other modifications included installing heavy metal plates to protect the aircraft belly from debris during tests, two 100-gallon water tanks for fire suppression, high-speed video and film cameras to record tire reactions, and over 200 sensors to record loads, pressures, temperatures, slip angles, and other critical data.

Over 1,000 individual new parts were fabricated, most by the shops at Dryden. The modifications added more than 40,000 pounds to the aircraft's empty weight. The LSRA modifications were completed in the fall of 1992. Following extensive ground tests, functional flights were flown to verify that all aircraft systems operated properly. Checkout of the landing gear test system continued throughout the summer of 1993.

In the fall of 1993, the LSRA flew to the Kennedy Space Center and performed about 25 test missions, landing on Kennedy's 15,000 foot runway. Tests consisted of a landing on the LSRA's main landing gear, establishing the desired test conditions, then forcing the test gear down in a precise manner. Test results will not only help develop better landing gears, but also to develop landing techniques that will wrestle maximum performance from the landing gear yet help prevent over-stressing any part of the system.

The LSRA remained active in 1995 and is available to industry to assist in the development of commercial landing gears, wheels, tires, and brakes.

A close-up view of the LSRA shuttle landing gear installation. (NASA Photo)

PART II: TESTBEDS

C-130 RAMTIP Program Testbed

AIRCRAFT TYPE: C-130 Hercules
MISSION: development of advanced cockpits
TIME PERIOD: Early 1990s
RESPONSIBLE AGENCY: Air Force Systems Command/Air Force Logistics Command
CONTRACTORS: Lockheed Aeronautical Systems, Bendix-King Avionics, GEC Avionics, Honeywell Inc, Smith Industries, and Lockheed Sanders

THE STORY

In 1991, a testbed C-130 was modified with the so-called $16 million Reliability and Maintainability Technology Insertion Program (RAMTIP) cockpit. The high-tech system used flat-panel, liquid crystal displays (LCDs) as a replacement for the mechanical instrument panels at the pilot, copilot, and navigator stations. The LCDs were the largest ever used on a C-130 at the time.

The engineering tasks under this project were to design, develop, fabricate, install, demonstrate, and support an engineering prototype of a true electronic C-130 cockpit. The engineers were looking into the operational effectiveness and suitability of liquid crystal, flat-panel displays and verification of projected improvements in the LCD's reliability, maintainability, and economy as compared to the standard electromechanical instruments and cathode ray tubes (CRTs).

Six, full-color, active-matrix displays – each measuring six by eight inches – replaced more than 60 mechanical-type cockpit instruments on the standard C-130. Five Displays were on the main instrument panel and one was at the navigation station.

The displays presented flight attitude, navigation, weather, radar and engine-operating data on the screens. The LCDs used digital processing, which required reduced electrical power, weighed less, and were more durable than CRTs.

The LCDs operate using the unique characteristics of liquid crystals whose properties are between those of a crystalline solid – whose properties vary with direction – and a liquid, whose properties are the same in all directions.

"The LCD is expected to provide us with the next generation of electronic displays for military aircraft, succeeding the CRT," explained Jerry Cazzell, the RAMTIP Program Manager.

Compared to CRTs, the new displays have been described as offering savings of about 60 percent in volume, 70 percent in weight, and 80 percent in power. With increased production capability, LCD costs may be only 50 percent of the cost of the CRT, and the LCD has been proven to be 10 times more reliable. LCDs also have the advantage over CRTs of "graceful degration," i.e., only parts of the LCD screen fail – not the entire display as is often the case when a CRT goes down.

The first flight of the testbed C-130 with the new LCD cockpit took place in February 1991 at Dobbins Air Force Base, Georgia and lasted just over four hours. After additional flight tests, the aircraft was returned to the 314th Tactical Airlift Wing at Little Rock AFB, Arkansas. The aircraft was then tested and maintained by the Military Airlift Command (MAC) during the test and evaluation period.

Following completion of flight testing the aircraft was retired. Although the test results showed the advantages of LCD technology when applied fleet-wide, continuing to operate one uniquely-modified aircraft proved too expensive. The modification was so massive that returning the aircraft to a standard configuration also was too expensive and it was removed from service.

Standard C-130 instrument panel appears to be a montage of instruments compared to the RAMTIP cockpit design. (USAF Photo)

The actual RAMTIP cockpit. (USAF Photo)

ONE-OF-A-KIND RESEARCH AIRCRAFT

Falcon ATLAS Program Testbed

AIRCRAFT TYPE: Falcon 20 Executive Jet
MISSION: Development of high-tech guidance system
OPERATING PERIOD: 1991 to mid-1990s
RESPONSIBLE AGENCY: Wright Laboratory Armament Directorate, Eglin Air Force Base, Florida

THE STORY

The name of this program, ATLAS, stands for Advanced Technology LADAR System, which was carried out at Eglin Air Force Base during the early years of the 1990s. The goal of the program was to develop guidance systems which would allow weapon delivery from a safer distances.

Such a system would allow the launching aircraft to be free of the battlefield environment, away from anti-aircraft artillery and surface-to-air missiles. The hoped-for guidance system would then search the ground for targets, recognize the targets and then track the missile to impact. Meanwhile, the aircraft would either be on its way back to base or off to another target.

The guidance system components were carried aloft for testing activities by a modified Falcon-20 executive jet testbed in a pod mounted on its under-fuselage. The testbed aircraft also carried a number of on-board systems to collect data as the ATLAS guidance system looked at targets while the aircraft was in flight.

The ATLAS system consisted of a LADAR sensor, a model-based automatic target recognition system processor, and a Global Positioning System-aided inertial guidance system.

The LADAR (for laser radar) worked in a similar manner to a law enforcement radar gun to catch speeders. The difference in the LADAR system was that the system used a laser light beam instead of an electronic signal.

In the ATLAS guidance system, the laser scanned the ground beneath the weapon, creating a high resolution image of the target area. From that data, a three-dimensional shape map was created from the data.

Next, the target recognition processor, which was programmed with the shape of the target then compared the shapes of the image with pre-programmed information searching for the match. The system then had the capability to guide the weapon to the target.

Eglin engineers felt that the program could have real application to both fixed and mobile targets, which would give the pilot greater flexibility with his weapons.

Besides its obvious advantages for aiding in weapon delivery, the ATLAS system also showed a number of possible applications for commercial use. It was considered extremely feasible for use in housing and urban development because of its ability to scan, store, and process extremely accurate three-dimensional information.

The ATLAS pod appears similar to a belly-mounted external fuel tank. (USAF photo)

PART II: TESTBEDS

F-4 Fly-By-Wire Program Testbed

AIRCRAFT TYPE: F-4 Phantom (In Air Force Museum)
MISSION: Development of electrical flight control systems
OPERATING PERIOD: 1969 to 1974
RESPONSIBLE AGENCY: Air Force Flight Dynamics Laboratory
CONTRACTOR: McDonnell Douglas

THE STORY
This particular McDonnell-Douglas F-4 Phantom (62-12200) exemplifies the context of this book. This aircraft has been modified, and remodified many, many times for a number of test and development programs. The mere mention of the numbers 12200, and most people in the test business would know this aircraft.

USAF Prototype Fighter Programs (YRF-4C and YF-4E)
This unique Phantom, number 266 off the production line, was originally scheduled as a Navy F-4B. Instead, it was modified into a USAF prototype aircraft when the Air Force began using the famous fighter type. It first became the YRF-4C prototype for a line of USAF reconnaissance planes which would eventually number several hundred. When it completed tests as the YRF, this versatile Phantom would be returned to the modification hanger to be converted into the YF-4E, the prototype for more than a thousand F-4E and F-4F fighter-bombers.

In that latter configuration, the Phantom "chin" carried a 20mm rapid-fire cannon, rather than a camera, and marked the beginning of a series of gun-equipped F-4s that continued into quantity production.

Agile Eagle Program
It was then time for this versatile F-4 to move from prototype work into its next life as a testbed aircraft. The Agile Eagle program involved the installation of leading edge wing slats to test concepts for improved maneuverability. The fighter performed well, and the slats ended up being included on all F-4 fighters that subsequently were deployed into the Air Force fleet.

SFCS Program
In 1969, with its camouflage replaced with a colorful blue and white paint scheme, 12200 became the testbed aircraft for the Survivable Flight Control System (SFCS) Program which addressed the development of fly-by-wire technology. While previous control programs (B-52 LAMS and CCV) addressed large flexible aircraft, the SFCS Program was directed specifically toward future fighter aircraft.

Using a fly-by-wire system, the pilot could "command" the aircraft with greatly-reduced physical effort through a system of

This F-4 testbed aircraft started life as the YRF-4 prototype before being converted into a number of other flight control test configurations. (USAF Photo)

ONE-OF-A-KIND RESEARCH AIRCRAFT

The venerable testbed carried the "Fly-By-Wire" nomenclature throughout its many programs. (Bill Holder Photo)

The PACT (Precision Aircraft Control Technology) Program was directed toward improvement in combat maneuvers in future fighter aircraft. (Bill Holder Photo)

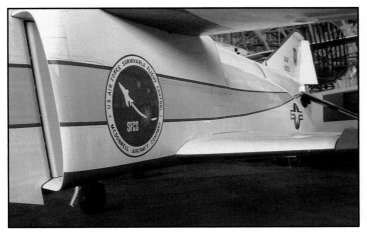

The Survivable Flight Control System (SFCS) modification to the F-4 testbed addressed the development of fly-by-wire technology. The testbed was characterized by flashy full fuselage-length "racing stripes." (Bill Holder Photo)

The venerable testbed, possibly the most-modified testbed ever, now resides in a position of honor at the Air Force Museum. It's certainly deserved. (Bill Holder Photo)

The Fly-By-Wire F-4 testbed flairs out on a final approach at Edwards Air Force Base, California. (USAF Photo)

electrical wires, rather than conventional complex mechanical systems using pulleys, linkages, and pushrods. Thus, the pilot could fly more smoothly and precisely with less effort devoted to flying the aircraft, enabling him to concentrate on other important tasks.

In the design of the SFCS, aircraft motion, rather than the control surface was the parameter to be controlled by pilot inputs. SFCS was also the first USAF aircraft to fly with a totally new fly-by-wire system having no mechanical links between stick and control surfaces. Sidestick controllers were installed for evaluation in conjunction with the improved handling qualities. Much of the SFCS configuration was derived from the earlier TWEAD (Tactical WEApon Delivery) Program.

The SFCS was evaluated with a test program of 84 flights from Edwards Air Force Base and McDonnell Douglas in St. Louis.

During the test program, the SFCS F-4 demonstrated that the new system made possible better flight control, improved safety, greater maneuverability and dramatic changes in aircraft designs of the future.

PACT Program

The final test mission for 12200 was being the main player in the so-called Precision Aircraft Control Technology (PACT) program. This program was started as an Air Force Flight Dynamics Laboratory program, but during the program, the Air Force efforts were re-directed to the F-16 CCV Program. McDonnell Douglas, though, continued the program with company funding and retained the same PACT name.

Airframe changes to the now-venerable F-4 were quite evident with the installation of two closely-coupled front canard surfaces, each of which had an area of 20 square feet. The canards were designed such that their outer panels could be removed, leaving each canard with an area of 8.5 square feet.

The PACT configuration paved the way for demonstrated improvements in combat maneuvers. Finally, in 1979, it was time to retire 12200, and what better place to recognize its accomplishments than to install it in the Air Force Museum where it remains to this day. Its final flight wasn't under its own power, but suspended in a sling under a helicopter for the trip from St. Louis to Dayton.

ONE-OF-A-KIND RESEARCH AIRCRAFT

F5D Skylancer Hypersonic Glider Testbed

TYPE OF AIRCRAFT: F5D Skylancer
MISSION: Simulate X-20 DynaSoar maneuvers
TIME PERIOD: 1960s
RESPONSIBLE AGENCY: NASA Dryden Flight Research Center

THE STORY

It was an exciting time for space development in the early 1960s, and testbed aircraft were playing a big part in the effort. In this particular case it was the X-20 DynaSoar program.

A NASA F5D Skylancer testbed played an important part with simulations of off-the-pad escape and landing maneuvers. This particular testbed was used because of its low lift/drag (L/D) ratio and this resemblance of its planform to that of the X-20 planform.

The proposed hypersonic glider would have been launched vertically from a large booster rocket and landed unpowered. Flight crew safety concerns in the event of a booster malfunction on the pad, or shortly after launch led to the proposal of an auxiliary booster to pull the glider away from the danger area so that the pilot could assume control and land nearby. However, such hypersonic gliders had low L/D and were landed unpowered. In addition, thermal-structural considerations led, then as now, to minimally-sized windows, limiting the pilot's field of view.

The F5D testbed aircraft simulated the escape maneuvers which were initiated with a high-speed run approximately a thousand feet above the ground. The pilot then pulled up vertically and cut the power, and then extended the speed brakes. This maneuver simulated the auxiliary-booster-rocket burnout.

The approach and landing maneuvers examined were 360 degree-spiral and straight-in approaches. A blue-amber system was used to restrict the visibility, with two different window configurations being examined.

This is one of two F5D-1 Skylancers modified by NASA for research operations. The aircraft was involved with simulations of DynaSoar landings. (NASA Photo)

The blue-amber system used a transparent blue visor with a transparent amber plastic lining of the cockpit. With the visor up, the pilot could see outside the aircraft. With the visor down, he could still see the instruments, but couldn't see out of the canopy because amber was the complementary color to blue.

The simulated escape maneuvers were acceptable to the pilots, with good control achieved. The circular pattern was again preferred, and flare control was not affected by the restricted visibility. The visibility restriction did not interfere with the navigation capability, although it did adversely affect portions of the escape maneuvers and landing approaches at times.

After the DynaSoar Program was canceled in the early 1960s, NASA Dryden continued to fly the F5Ds in support of lifting body and SST studies, retiring the aircraft in 1970. Incidently, Neil Armstrong was one of the test pilots during the program and the plane currently rests in front of the Neil Armstrong Museum in western Ohio.

This particular F-5D, which was flown by Neil Armstrong, and is now on display in front of the Neil Armstrong Museum at Wapakoneta, Ohio, just north of Dayton. (Phil Kunz Photo)

PART II: TESTBEDS

F-8 Supercritical Wing Testbed

AIRCRAFT TYPE: Vought F-8
MISSION: Aerodynamic Research
OPERATING PERIOD: 1971-1973
RESPONSIBLE AGENCY: NASA Dryden Flight Research Center

THE STORY

In the 1960s, Dr. Richard T. Whitcomb of NASA's Langley Research Center in Hampton, Virginia, developed an airfoil shape that promised greatly reduced drag at high subsonic speeds. Aerodynamicists long knew that as the speed of an aircraft reached about Mach .8, or about 530 miles per hour, small regions of supersonic flow developed around the aircraft. This speed was called the aircraft's critical Mach number. These areas of local supersonic flow occurred particularly on the top of the wing. The resulting shock wave caused turbulent airflow to develop, greatly increasing drag, causing buffet when the turbulent air hit the tail, and changing the aircraft's stability characteristics. Dr. Whitcomb theorized that if the point where the shock wave developed could be moved further aft on the wing, then the area of turbulent flow would decrease, as would all the other undesirable effects that occurred at the critical Mach number. In addition, with smooth air flowing over more of the wing, more lift could be produced, further increasing efficiency.

The most widely used means to reduce drag at the critical Mach number had been to sweep the wings. However, this carried the penalty of increased structural weight and poor low speed handling characteristics . . . sometimes tolerable on military aircraft, but always undesirable on airliners. Dr. Whitcomb's airfoil consisted of a mostly flat top, and a convex bottom (as on most other airfoils) but ending in a concave area near the trailing edge. Wind tunnel studies verified that a wing using this airfoil shape could push the critical Mach number to .95. Thus the shape became known as the supercritical wing.

The potential for this wing was obvious. Future aircraft could fly at higher speed but at the same power as present aircraft. Conversely, they could fly at the same speed but at reduced power. This translates into the flexibility of lower operating costs, increased operating range, or increased payload.

By 1969, NASA Dryden developed a program to flight test an aircraft with a supercritical wing in order to verify the theory and wind tunnel predictions. The U.S. Navy provided an F-8 Crusader. NASA engineers developed a 43 foot-long wing shape representative of an airliner wing. By September 1969, a contract to construct the wing was awarded to the Los Angeles division of North American Rockwell.

The completed supercritical wing was (SCW) delivered to NASA Dryden in late 1970 and installed on the F-8, which now carried the designation NASA 810 and became known as the SCW F-8. The aircraft made its first functional check flight in March 1971.

The SCW F-8 flew a series of preliminary checks and evaluations between March and May, achieving a top speed of 725 mph and a peak altitude of 46,000 feet. No problems with the wing were observed. Following these flights, approximately 240 pressure sensors, installed when the wing was built, were activated for use on the next series of flights. The purpose of these sensors was to determine the exact location where the shock wave formed.

During the summer of 1972, side fairings were added to the fore and aft sections of the fuselage. These produced the "Coke-bottle" shape reminiscent of aircraft such as the F-105 and T-38. More formally known as area ruling, the fairings reduce drag by keeping the cross section area of the wing and fuselage nearly constant. This principal had been developed in the 1950s, also by Dr. Whitcomb. The purpose of adding them to the F-8 SCW was to determine if the combination of area ruling and the supercritical wing increased efficiency greater than either acting alone.

Flight testing concluded by May of 1973, with over 75 flights being performed. The efficiency improvement predicted by analysis and wind tunnel testing was demonstrated. The technology was quickly transitioned to the airline industry and incorporated into new aircraft being designed.

Its mission complete, the F-8 SCW now sat idle in the hangar at NASA Dryden. No other research programs for the aircraft ever came to fruition. In August 1991, it was sent to a contractor for restoration, and in June 1992 it was placed on static display at NASA Dryden, alongside the Digital Fly-by-Wire F-8.

The F-8 Supercritical Wing test aircraft in flight. The new wing had a typical airliner-type platform. Note also the fuselage mounted fairings to reduce drag. (NASA Photo)

The F-8 testbed is now on display at NASA Dryden at Edwards Air Force Base. (Steve Markman Photo)

ONE-OF-A-KIND RESEARCH AIRCRAFT

F-8 Digital Fly-By-Wire Testbed

AIRCRAFT TYPE: Vought F-8 Crusader
MISSION: Digital Flight Control Research
OPERATING PERIOD: 1972-1985
RESPONSIBLE AGENCY: NASA Dryden Flight Research Center

THE STORY

The NASA Digital-Fly-by-Wire (DFBW) F-8 was developed and operated to investigate the use of digital computers and fly-by-wire technologies to control high performance aircraft. As modern civilian and military aircraft flew faster and higher, and carried more varied payloads over longer ranges, the flight control systems were being tasked to perform more and varied tasks in order to keep the aircraft flight characteristics acceptable to the pilot.

By the early 1970s, analog computers were already helping to accomplish this by augmenting the basic mechanical controls. Fly-by-wire offered the potential to eliminate the heavy and maintenance-intensive mechanical push rods, cables, and levers by replacing them with electrical wires. Digital computers offered greater flexibility and reliability than analog computers for performing the needed computations.

Fly-by-wire technology also offered another advantage. Because the control surfaces would always be active, even when not commanded by the pilot, the aircraft no longer had to be aerodynamically stable. The computer would constantly send small commands to each control surface not just when maneuvering, but even to keep the aircraft straight and level. The end result was an artificial stability in that the aircraft felt stable to the pilot. This could allow smaller horizontal and vertical stabilizer surfaces, with resulting reduced weight and drag. Air Force studies indicated that future aircraft could be 20 percent lighter yet less expensive using this technology.

Aircraft such as the Concorde offered a "pseudo fly-by-wire" that allowed the pilot to control the aircraft through the autopilot, but still retained mechanical controls. The DFBW F-8 would be the first test aircraft to demonstrate the advantages of digital fly-by-wire and also to eliminate the mechanical system, even as a back-up.

By 1971, NASA's Dryden Flight Research Center selected and began modifying an F-8 Crusader, a supersonic Navy jet fighter. A digital computer, inertial measuring unit (which measures the aircraft attitude), and other equipment originally built for the Apollo Lunar Module were incorporated into the modification. A three-channel analog system served as a backup to the digital fly-by-wire system. Initially, the mechanical flight control system was retained until the reliability of the fly-by-wire system could be verified. A second F-8 was obtained and modified to serve as a ground simulator and "iron bird." It used actual flight hardware and served to check the system design and integration into the aircraft systems.

Modifications were completed in early 1972. The DFBW F-8 carried the designation NASA 802 and a distinctive lightning bolt ran down the side of the white fuselage. Following ground checkout, it made its first flight in May 1972.

Following a year of checkout flights, a test of the potential effects of lightning strikes was performed. Since a digital fly-by-wire system is completely dependent on the reliable operation of black boxes and wires, the system must be protected and completely immune to damage by lightning. A mobile test facility was built at NASA Dryden to produce artificial lightning bolts that would travel through the F-8. A generator that could produce lightning bolts of up to 1000 amps was mounted near the nose. Electrical equipment in the aircraft was attached to test equipment so researchers could monitor their operation during the strikes. Tests were run with the strikes running from nose to tail, nose to wingtip, and wingtip to wingtip.

Forty-two flights were ultimately flown in the first phase of flight testing, concluding in the summer of 1973. During these flights, the mechanical system eventually was removed to demonstrate complete reliance on the fly-by-wire system. These flights paved the way for the Space Shuttle Orbiter which would make use

The F-8 Fly-By-Wire testbed was one of a long list of aircraft investigating this unique technology. (NASA Photo)

PART II: TESTBEDS

Two unique F-8 testbeds. The F-8 Fly-By-Wire testbed flies in formation with its F-8 brother, the supercritical wing testbed. (NASA Photo)

of this technology. They also tested a prototype of the side stick that would be used in the F-16.

The flights performed in the first phase used a single digital computer. Although the system proved reliable, future production aircraft would need even greater redundancy. During the next three years, the single Apollo-vintage computer was replaced with a triplex digital system. The triplex analog backup system was retained. Approximately 30 flights were flown between 1976 and 1978. One of the objectives of this phase was to verify that the computers could monitor themselves, detect an error in any one, and switch the failed computer out of the system.

Following this effort, further research in digital fly-by-wire technology continued. A program called Analytical Redundancy Management Test Technique was performed in 1979 and 1980. The purpose of this effort was to show improved reliability with a reduction in the number of sensors on the aircraft. Sensors tell the digital flight control computer the states of the aircraft, i.e. its attitude, speed, altitude, acceleration, etc. Each of these sensors had to be redundant to protect the system in the event of a failure of any one of them. Software in the digital flight control computer could detect failures and select alternate actions. The goal was to demonstrate a potential decrease in the number of sensors by 25 percent while maintaining acceptable levels of reliability.

In 1981, the F-8 was used to help fix a problem with the Space Shuttle Orbiter. The Shuttle had encountered a problem called a pilot induced oscillation, or PIO. This problem can be encountered when the aircraft does not respond as quickly as the pilot thinks it should. When the pilot does not feel the aircraft responding, he increases his inputs to the controls . . . just as the aircraft starts responding to the original command. This results in too much motion, so the pilot takes out some command. When he doesn't feel the change in motion, he takes more out . . . just as the aircraft starts responding to the original correcting command. Now he has too much motion again but now in the other direction. The result is an oscillation that can get out of control quickly. The solution is for the pilot to let go of the stick and let the aircraft's natural stability dampen the motion. But this is easier said than done with an out-of-control aircraft only a few feet off the runway or while in close formation with another aircraft.

The solution, tested on the F-8, was an adaptive filter, commonly called a PIO suppressor. The computer keeps track of typical pilot control inputs. When it senses a shift from gentle, gradual inputs to the quick, almost microscopic inputs typical of a pilot about to enter a PIO, the computer makes the aircraft respond less to any given command. In essence, the pilot's command is filtered, or suppressed, until his commands become more normal. This gives time for the aircraft's own stability to dampen the motion. This technique worked well enough that it was incorporated into the Shuttle.

Another research program flown in 1981 was sensor redundancy tests. Given that most aircraft motions occur on several axes at once and are interrelated, many of them can be derived mathematically from the values of other known parameters. Thus, if there are ten sensors measuring ten different parameters and one of them fails, the theoretical value from the failed sensor can be calculated by knowing the values from the remaining nine! This technique had potential for decreasing the number of sensors needed to provide acceptable reliability.

Potential solutions to generic software problems were tested in 1982. A generic problem is one in which each computer, running identical software, can produce the same error. Despite extensive testing, these shortcomings in the software sometimes go undetected until the aircraft encounters some unusual or unlikely combination of conditions that had not been anticipated. Then, the aircraft may not respond properly even though all the computers are working properly and giving the same answer. The solution to such a generic problem is to write separate software for each of the computers in the system. The software function is identical, and for a given input, each set of software will produce the identical answer. However, each solution will be produced by the computations having been arranged differently and having been written by different programmers. An undiscovered fault in one set of software would be very unlikely to occur at the identical place in the software of the other computers. Flight tests proved this technique for protection against generic faults was very effective.

A related program also was flown in 1984. This time, each computer contained its own set of backup software that functioned identically. Normally, when one computer gives an answer different from the others, it is assumed the computer has failed and is switched out. In this latest test, a software error was first assumed, rather than a hardware error, and the errant computer switched to its backup software. If this did not correct the error, the computer was then switched out. All this comparison and switching occurred automatically without intervention by the pilot.

The DFBW F-8 made its last flight in 1985. It then sat quietly inside a NASA Dryden hangar, awaiting the chance to fly other programs. None ever materialized. In May 1992, the DFBW F-8 was placed on static display in front of NASA Dryden, alongside the Super Critical Wing F-8.

The F-8 Fly-By-Wire testbed aircraft is now on display at NASA Dryden at Edwards Air Force Base. (Steve Markman Photo)

PART II: TESTBEDS

F-15 AECS Program Testbed

AIRCRAFT TYPE: F-15A
MISSION: Testbed aircraft for advanced environmental system development
OPERATING PERIOD: Mid-1970s
RESPONSIBLE AGENCY: Air Force Flight Dynamics Laboratory
CONTRACTOR: McDonnell Douglas

THE STORY

An Advanced Environmental Control System (AECS) was developed by the Air Force Flight Dynamics Laboratory (AFFDL) during the 1970s. An F-15A was the testbed aircraft for testing the system.

A flying air conditioner of sorts, the AECS was designed to cool aircraft electronics which sometimes overheat during high performance flight. The system could also give the pilot a more comfortable cockpit during take-off, flight, and landing.

The AECS F-15A was modified at the McDonnell Douglas St. Louis facility, and performed its first flight test in May 1977.

The AECS program, though, actually began life three years earlier because of the rising costs of repairing avionics that failed due to overheating.

Flight Dynamics Lab engineers explained that when aircraft electronics reached over 100 degrees, small components such as transistors, resistors, and diodes just don't work. As the reliability of avionics systems went down, so did mission reliability, and maintenance costs greatly increased.

It was explained that about 52 percent of all avionics system failures were related to environmental conditions like temperature, humidity, and dust. Of those environmental-related failures, 55 percent were linked to high temperatures.

The AECS was a regenerative system that used on-board heat sinks to dispose of hot air. In contract, most environmental control systems dumped heated air into the atmosphere, causing severe drag penalties on the aircraft's performance at high speeds. Specifically, the AECS heat sinks were the fuel and discharged or "used" air from the cockpit and avionics bays.

A particularly unique portion of the ACES was the turbo-compressor which was equipped with automatically-adjustable vanes which allowed greater cooling capacity at all flight regimes.

The AECS used a commercially available dust separator for cleaning the air and a high-pressure water separator to keep the humidity low. Because of these unique machines and cooling techniques, the AECS could deliver clean, dry air at 40 degrees F into the aircraft's avionics bay and cockpit at all times.

All of the AECS flight testing was accomplished during 1977 at both Edwards Air Force Base, California, and Chase Field in Beeville, Texas.

ONE-OF-A-KIND RESEARCH AIRCRAFT

F-15 ASAT Program Testbed

AIRCRAFT TYPE: F-15A
MISSION: Demonstration of the launching of an anti-satellite weapon system from an F-15 launch aircraft
OPERATING PERIOD: Mid-1980s
RESPONSIBLE AGENCY: U.S. Air Force
CONTRACTOR: McDonnell Douglas

THE STORY

Development of an anti-satellite weapon system has long been a goal of the Department of Defense, and a number of ambitious and unsuccessful efforts have taken place since the 1960s. Most of the efforts were extremely complicated, but a 1980s program using a fighter-launched missile proved that it could be done.

The ASAT program actually began earlier as an outgrowth of the Prototype Miniature Air Launched Segment (PMALS) anti-satellite program. As the program matured, it was changed from PMALS to the more descriptive ASAT nomenclature.

The program involved the modification of an F-15A giving it the capability to carry and launch a 2,700 pound two-stage missile. The missile's first stage was from an AGM-69 Short Range Attack Missile (SRAM), with a second stage consisting of a 6,000 pound thrust Altair III Thiokol rocket motor. Other features on the system included a seeker system and small rocket motors to aid on keeping the missile on course.

A single live-fire test of the system was accomplished on Sep-

The F-15 ASAT Program was identified on the F-15A testbed aircraft by this characteristic logo on the tail. (USAF Photo)

This is definitely a unique space missile launching platform. The concept proved successful knocking down a U.S. satellite, but the system would never be fielded operationally. (USAF Photo)

PART II: TESTBEDS

The F-15A ASAT testbed aircraft is seen being mid-air refueled by a KC-135 tanker. The ASAT weapon can be seen mounted beneath the aircraft's fuselage. (USAF Photo)

tember 13, 1985 when the F-15 was flown at a nearly vertical path to an altitude of 80,000 feet and then launched the missile. The particular F-15 used for this program was one of several based at Edwards Air Force Base for various test support missions. The ASAT destroyed a U.S. Solwind P-78-1 Gamma Ray Spectrometer satellite which was used as the target. Despite the success of the launch, the test was never repeated and the unique ASAT system, developed to counter suspected Soviet satellites, was never continued.

The program was divided into two phases, the modifications to the launching F-15A testbed and the development of the weapon system itself.

Modifications to the aircraft, which were known as Group A, included the installation of a modified central computer, expanded environmental control system (ECS) ducting, reinforced ammunition bay cover, and ammunition conveyer restraint.

The Group B missile-associated components included the two-stage ASAT missile, dedicated centerline pylon, modified ammo door and carrier aircraft equipment (C&E) pallet.

The launch technique used the so-called "Zoom" launch maneuver with the launching aircraft providing a significant velocity boost to the missile before its launch. The fact that the missile was launched at an extremely high altitude also greatly reduced the slowing aerodynamic drag forces the missile would have experienced at lower altitudes.

The F-15 testbed used for this program remained active at Edwards Air Force Base as of 1994.

ONE-OF-A-KIND RESEARCH AIRCRAFT

F-15 IFFC/ABICS/ICAAS Program Testbed

AIRCRAFT TYPE: F-15B
MISSION: To demonstrate the blending together of new technologies to improve a fighter's survivability and accuracy
RESPONSIBLE AGENCY/S: Air Force Avionics and Flight Dynamics Laboratory, U.S. Navy, Department of Defense, others
TIME PERIOD: 1970s and 1980s
CONTRACTORS: McDonnell Douglas, Martin Marietta, and Northrop Corp.

THE STORY

IFFC Program

It was called the Integrated Flight and Fire Control (IFFC)/ Firefly Program and had the purpose of demonstrating a technology which could potentially improve the combat effectiveness of all tactical aircraft. The IFFC design involved the blending of the operation of the flight control system and fire control system. Accurate delivery of air-to-ground ordnance during evasive maneuvering and pilot-aided air-to-air tracking were the cornerstones of the IFFC concept.

Delivering ordnance on a ground target during a turning maneuver was the goal of the air-to-ground portion of the IFFC program. The system tried to eliminate the conventional manner of rolling out to a wings-level, stable attitude before releasing the weapons which always left the aircraft vulnerable. The IFFC program goal was to accomplish the release under varying maneuvres and at high speeds.

The two-seat F-15B, SN 77-166, was selected as the testbed aircraft for this joint Air Force Avionics and Flight Dynamics Laboratory program. The modifications to convert this F-15 into the IFFC configuration were the minimum possible in order to use as much off-the-shelf hardware as possible.

The pilot was intended from the very beginning to be a heavy player in this program, ranging from full manual control to pilot-aided automatic control. Also, with the IFFC configuration, the pilot was able to designate particular axes for automatic control while retaining manual control of the remaining axes. The system was also designed to aid the pilot in the terminal phase of the attack mode where precision control was critical.

The IFFC system addressed improvements in both air-to-air and air-to-ground operations. With air-to-air, the system delegated the control authority to an automatic system. Then, the pilot's role was simply to keep the target within the confines of a sighting rectangle whose dimensions corresponded to the authority limits of the automatic system. Then the IFFC system accomplished the firing solution.

A major part of the IFFC system was the so-called Firefly 3, a General Electric aircraft concept, which included an electro-optical sensor/tracker pod, a data processor, and a fire control system using target data generated by the system.

This F-15B was initially modified to the IFFC configuration. (McDonnell-Douglas Photo)

PART II: TESTBEDS

There was no mistaking the IFFC-configured F-15B which carried this distinct logo on the tail. (McDonnell-Douglas Photo)

ONE-OF-A-KIND RESEARCH AIRCRAFT

The primary modification to the IFFC F-15B was the addition of the Martin-Marietta ATLAS II (Automatic Tracking Laser Illumination System) which served as the primary sensor and tracker. The ATLAS II pod was mounted on the underside of the left engine nacelle.

The flight test program, consisting of some 200 flight tests, took place at Edwards Air Force Base in 1981. The ambitious goals for air-to-ground gunnery and bombing were a 10-1 increase in survivability against anti-aircraft artillery, and a 2:1 increase in weapon delivery accuracy over conventional methods.

Much of the data acquired from the IFFC program was funneled directly into the AFTI/F-16 Program.

ABICS Program

Following completion of the IFFC program, the same 77-166 F-15 was the main player in the ABICS program. The nickname stood for Ada Based Integrated Control System, which tested the effectiveness of the Ada software. Both 16 and 32 bit Ada Computers were flight tested in this program.

The purpose of the program, which fitted the testbed aircraft with both new software and hardware, was the development and flight demonstration of an inertial reference assembly and the Ada flight control system.

Externally, the ABICS configuration didn't appear any different from a fleet aircraft. The fuselage-mounted pod, that was a part of the IFFC configuration, was removed for the ABICS mission.

Through the course of the ABICS program, there were a number of different versions including ABICS I, II, and III which examined various flight and fire control techniques. An ABICS IV was planned, but never reached fruition.

The ABICS configuration was first flown in September 1984, culminating with integrated flight, fire, and navigation systems using multi-function sensor testing carried out in March 1987. The system was evaluated by 20 pilots in 52 flights totalling some 70 flight hours.

ICAAS Program

The final configuration that 77-166 would experience would be the ICAAS (Integrated Control & Avionics for Air Superiority) pro-

The final configuration of this testbed with the so-called ICAAS program. The configuration could not be identified externally by any distinctive markings. The ICAAS program was canceled before completion. (USAF Photo)

gram. The purpose of this program was to develop technology to kill and survive when outnumbered in an air combat scenario.

The ICAAS modifications consisted of a new glass cockpit, helmet display, avionics update, new power supply, updated central computer, and a modified radar. Unfortunately, the ICAAS aircraft saw only four flights before premature program termination.

The same 77-166 F-15B was modified from the IFFC configuration to the ABICS configuration. The IFFC tail identification was removed and replaced with the original Edwards tail marking. (USAF Photo)

PART II: TESTBEDS

F-15 Highly Integrated Digital Electronic Control Testbed

AIRCRAFT TYPE: F-15A
MISSION: Develop electronic flight and engine control system
TIME PERIOD: 1990s
RESPONSIBLE AGENCY: NASA Dryden Flight Research Center
CONTRACTOR: McDonnell-Douglas

THE STORY

The F-15 Eagle has proved itself in combat as the best aircraft in the world in the 1980s and well into the 1990s. The craft, though, would also prove itself as an excellent testbed aircraft and simulator having participated in a large number of test programs. This particular F-15A was acquired by NASA Dryden from the Air Force in 1976, and has participated in more than two dozen research projects through the years.

One of the more recent uses of this particular Eagle is the so-called HIDEC (Highly Integrated Digital Electronic Control) program which was carried out at NASA Dryden. The purpose of the program was to conduct flight research on integrated digital electronic flight and engine control systems.

The F-15 HIDEC testbed demonstrated improved rates of climb, fuel savings, and engine thrust by optimizing systems performance. The aircraft also tested and evaluated a computerized self-repairing control system for the Air Force that detected damaged or failed flight control surfaces. The system then switched to alternate control laws so undamaged flight surfaces could compensate for the damaged ones, allowing the mission to continue or the aircraft to land safely.

In one of its tests, the HIDEC testbed demonstrated the ability to be flown and landed safely using only engine power for control. The HIDEC flight control system had the capability to program the engines to turn, climb, descend, and eventually land the plane by varying the speed of the engines one at a time or together. This effort was inspired by the 1989 crash landing of an airliner in which the pilot maneuvered to a landing using engine power alone, following total loss of all control surface motion.

The F-15 HIDEC plane was highly-instrumented and equipped with an integrated digital propulsion-flight control system. It had the capability of being flown over a broad flight envelope to carry out its complex and sophisticated research projects. Also, the testbed carried an advanced version of the F100 EMD powerplant instead of the normal F100-PW-110 or 220 engine. With this powerplant, the HIDEC F-15 demonstrated a Mach 2 capability for its many test missions besides this specific HIDEC mission.

Other modifications made to the HIDEC aircraft included a cockpit panel with two thumbwheel controls, one for pitch and the other for banking turn commands. The system converted the pilot's inputs into engine throttle commands.

Nearly all the research which was carried out in the HIDEC program was applicable, according to NASA, to future civilian and military aircraft.

The final flight for the HIDEC program took place on October 29, 1993. Undoubtedly, though, this one-of-a-kind testbed will continue to perform more research and testing, supporting new programs for many years to come.

The HIDEC F-15 test aircraft was used to carry out flight research on integrated digital electronic flight control systems. (NASA Photo)

ONE-OF-A-KIND RESEARCH AIRCRAFT

F-15 STOL/MTD, ACTIVE Testbed

AIRCRAFT TYPE: F-15B
MISSION: Development of short take-off and landing and high angle-of-attack maneuver capabilities
TIME PERIOD: 1985 to mid-1990s
RESPONSIBLE AGENCY: Air Force Wright Laboratory, NASA
CONTRACTOR: McDonnell-Douglas Corporation, Pratt & Whitney

THE STORY

F-15 STOL/MTD Program

The goal of the F-15 STOL/MTD was to develop the capabilities for future fighters to land on wet, short, bomb-damaged runways. In addition, the testbed aircraft also demonstrated the capability to improve combat maneuverability. There was no mistaking this testbed aircraft with its flashy red, white, and blue paint job.

The aircraft, a highly-modified F-15B Eagle, participated in a 1985-to-1991 test program to validate its unique Pratt & Whitney-built two-dimensional thrust vectoring/reversing fighter nozzle configuration. In addition to its unique engine nozzles, the STOL/MTD testbed was also distinguished by moveable control canards mounted on the forward fuselage. Other modifications included integrated flight and propulsion controls, rough-field landing gear, and an advanced cockpit which greatly reduced the pilot workload, in spite of the complexity of the maneuvers to be performed.

The initial phase of the program began in September 1988 and recorded 43 flight tests with standard, circular nozzles. Then, the Air Force and contractor personnel installed and ground-tested the aircraft's thrust-vectoring, thrust-reversing nozzles which were designed and manufactured by Pratt and Whitney.

The first flight of the F-15 STOL/MTD with the new nozzles was May 10, 1989, from Lambert-St. Louis International Airport. On June 16, the craft was ferried to Edwards Air Force Base, California. Shortly thereafter, the F-15 STOL/MTD demonstrated and evaluated the thrust-vectoring feature of the new nozzles coupled with the integrated flight/propulsion control system of the aircraft.

Rough field taxi tests at Edwards were concluded in September 1989 with the plane's special landing gear that permits the aircraft to operate on rough-bomb-damaged runway surfaces. In December 1989, the STOL/MTD demonstrated its Autonomous Landing Guidance capability by making a precision night landing at Edwards, relying solely on internal sensors, with no runway lights or ground-based navigational aids.

Thrust-vectoring assisted takeoff tests in April 1990 demonstrated a 38 percent reduction in the length of runway used. Continued thrust-reversing flight tests led to the first F-15 STOL/MTD short landing on May 22, 1990. Using thrust-reversing and anti-skid autobraking for stopping after touchdown, the STOL/MTD landing in 1,650 feet.

In more recent flight test, the STOL/MTD expanded its in-flight thrust reversing envelope to Mach 1.6 at 40,000 feet. Assisted by thrust vectoring, the aircraft's pitch maneuverability at high angle-of-attack was demonstrated to be 110 percent better than an aircraft with only conventional controls. A flight control system modification in April 1991 lowered the landing distance to 1,370 feet, less than on-third that of a normal fighter.

Tactical Air Command pilots flew the STOL/MTD in October 1990 and recommended its Autonomous Landing Guidance System for retrofit on all F-15Es. The Navy also flew the testbed in 1991 to evaluate its speedhold landing control laws for carrier use. Finally, the infrared signature characteristics of the STOL/MTD were measured in-flight to see if the thrust vectoring/thrust reversing nozzles could potentially reduce an aircraft's vulnerability to infrared threats.

The program was completed at Edwards in August 1991 and the aircraft was placed in storage.

For its accomplishments, the program was honored with the 1990 Aerospace Laureate Award for outstanding contributions to aerospace technology.

There was no missing the patriotic paint scheme of the F-15 STOL/MTD testbed aircraft. (USAF Photo)

PART II: TESTBEDS

Detail of the aircraft's unique maneuverable propulsion system is visible in this overhead view. (McDonnell Douglas Photo)

This head-on view of the F-15 STOL/MTD provides a great view of the engine-mounted front canards which provide greater maneuverability capability for the aircraft. (McDonnell Douglas Photo)

The two-dimensional nozzles in full-open position. (McDonnell Douglas Photo)

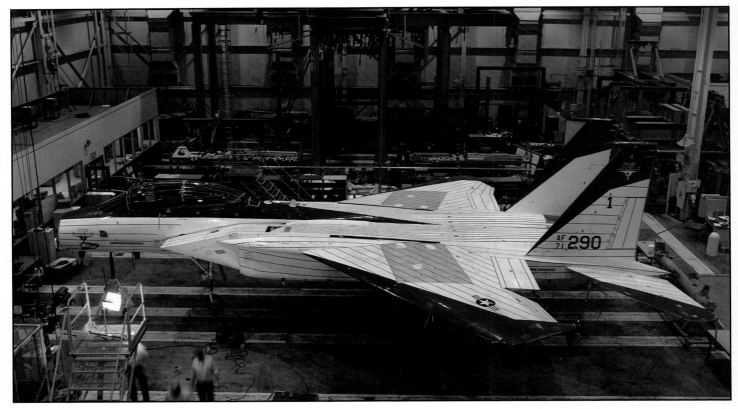

The F-15 STOL/MTD is shown under construction at the McDonnell Douglas facility at St. Louis. (McDonnell Douglas Photo)

F-15 ACTIVE Program

Following its use in the STOL/MTD Program, this particular sophisticated testbed aircraft could have languished in storage for many years. It's happened with other testbeds in the past. But there was another job to be performed and this highly-modified Eagle was brought back to life after 22 months of storage. The effort was the so-called ACTIVE Program, for Advanced Control Technology for Integrated Vehicles, which was a joint NASA/USAF/Pratt and Whitney effort.

The purpose of the program was to investigate symmetric pitch/yaw vectoring nozzle technology with an even more-advanced nozzle that had been fitted to the F-15. This nozzle, similar in concept to the one installed on the VISTA NF-16, could deflect thrust in any direction, not just up and down.

Wright Lab's Dennis Weiland indicated that development of such advanced nozzle technology could result in the use of nozzles for aircraft trim, thus reducing drag, both during maneuvering and cruise flight. Emphasis in the ACTIVE Program, though, concentrated on the latter interests.

For the ACTIVE Program, it was necessary to make some structural modifications for the aircraft to accept the new engine nozzles and fly with them. The airframe was also reinforced to accept the increased yaw forces, and the fuselage fairings were changed so that the vectoring engine nozzles would fit where the original engines had been located.

Undoubtedly in the years to come, additional advanced technologies will be tried and tested in this amazing F-15 testbed.

An artist's concept from early in the program showing how an operational F-15 STOL/MTD might look. (McDonnell Douglas Photo)

PART II: TESTBEDS

F-15 Streak Eagle Testbed

AIRCRAFT TYPE: F-15A
MISSION: To demonstrate performance capabilities of the F-15A
TIME PERIOD: 1972-1975
RESPONSIBLE AGENCY: USAF
CONTRACTOR: McDonnell Douglas

THE STORY

Without doubt, the F-15 has displayed great suitability for several test applications and programs. One of the most interesting applications was its use in establishing a number of climb to altitude records. It's using the "testbed" nomenclature in an entirely different context than is usually expected.

One of the first special-purpose F-15s was used in the so-called Streak Eagle program. The 'speedy' name was really appropriate in this application. The particular F-15 used in this program was the 19th pre-production aircraft. The Streak Eagle was modified in order to attempt to establish new international time-to-climb records.

The Streak Eagle was trimmed down as much as possible with all unnecessary items – missiles, radar, cannon, tail hook, one generator, utility hydraulic system, flap and speed brake actuators – being removed. Additionally, another 40 pounds were saved by not painting the aircraft. No excess baggage for this bird!

For the attempt, however, some specialized equipment had to be added including the following: A restraint device replaced the standard tail hook, a noseboom with alpha and beta vanes to determine angle-of-attack and sideslip along with special battery packs and controls. Other special equipment included an over-the-shoulder camera, a G-meter, standby attitude gyro, equipment for verifying altitude, and ballast.

The empty Streak Eagle weighed nearly 2,800 pounds less than a production F-15. Fuel varied from 3,000 to 6,000 pounds since only enough was carried to complete each profile.

The flights were made in early 1975, but as early as 1973 McDonnell Douglas engineers were convinced that the aircraft could

This particular F-15 testbed was easily recognizable from the flashy logo it carried on both sides of its nose. (Bill Holder Photo)

shatter time-to-climb records held by the F-4 and MiG-25. Altitudes and times of the records set were:

    3,000 meters (9,843 feet) 27.57 seconds
    6,000 meters (19,685 feet) 39.33 seconds
    9,000 meters (29,685 feet) 48.86 seconds
    12,000 meters (39,370 feet)59.38 seconds
    15,000 meters (49,212 feet)77.02 seconds
    20,000 meters (65,617 feet) 122.94 seconds
    25,000 meters (82,021 feet) 161.02 seconds
    30,000 meters (98,425 feet) 207.80 seconds

To put these feats in perspective: a Boeing 727 takes more than 15 minutes to achieve a 9,000 meter cruising altitude while the Streak Eagle did it in less than one minute! The Streak Eagle is now on display at the Air Force Museum.

The F-15 Streak Eagle was a different type of testbed, because it was specifically designed to capture time-to-climb world records, which it did. (Bill Holder Photo)

ONE-OF-A-KIND RESEARCH AIRCRAFT

F-16 Advanced Fighter Technology Integration Testbed

AIRCRAFT TYPE: F-16
MISSION: Development and Integration of Advanced Technology Concepts
OPERATING PERIOD: 1982 - Present
RESPONSIBLE AGENCY: Wright-Patterson Air Force Base Ohio
PRIMARY CONTRACTOR: General Dynamics/ Lockheed

THE STORY

The development of new military aircraft is a very long and intricate process, often taking over a decade from the initial requirements specification until the first operational aircraft is in service. Typically, as soon as a new aircraft enters service, planners begin looking at its replacement. They must project their thinking ahead at least a decade, consider the threats that may be encountered, and develop the technology to counter these threats. It was to this end, to develop the technology for new fighter aircraft to be developed in the 1990s, that the Advanced Fighter Technology Integration, or AFTI, program was born.

The AFTI program began at the Air Force's Flight Dynamics Laboratory, now the Wright Laboratory, in the early 1970s. With the F-15 about to enter service and the F-16 in early design, AFTI was to develop the technologies for the next generation of tactical fighters. The program not only was to develop and demonstrate the technologies, but also to integrate them in one vehicle to show the synergistic effects. What his means is AFTI would show that using all these new features together in an integrated system, the combined advantage would be greater than the sum of the advantages of individual features working separately. It would show how ad-

This 1980 artist's concept shows the design of the AFTI/F-16 configuration which would follow in two years. The concept incorporated fly-by-wire controls, advanced displays, integrated flight control system, and weapons fire controls. (USAF Photo)

vanced technologies could make aircraft more maneuverable, allow them to perform more diversified missions, and make them more survivable, yet decrease the pilot's workload, despite the increased complexity.

The original AFTI concept called for a scratch-built aircraft. But as the program progressed through the 1970s, the realities of cost soon limited the choice to modifying an existing aircraft. The

The AFTI/F-16 is one of the most capable research aircraft ever constructed. Begun in the early 1980s time period, the program was still active in the mid-1990s. (USAF Photo)

PART II: TESTBEDS

F-15 and F-16 were the natural candidates. In the summer of 1977, contracts were awarded to study conceptual designs to modify an existing aircraft: one to McDonnell Douglas Corp to study an AFTI/F-15 and one to General Dynamics Corporation to study an AFTI/F-16. These studies were to explore how these aircraft could be modified to demonstrate the use of direct force control for weapon line pointing, integrated fire and flight control, cockpit advancements, and digital flight control.

The AFTI/F-16 design was selected and General Dynamics was awarded a development contract in December of 1978. A pre-production version of the F-16, serial number 75-750, was provided as the test vehicle.

As the design took shape, the AFTI/F-16 took on a new appearance. A set of vertical fins was mounted on the air inlet, one on either side of the nose wheel, and a dorsal housing was added on top of the fuselage in which additional electronic equipment would be installed. The cockpit was also modified with advanced displays and controls. The new capabilities incorporated included:

- Direct Lift and Pitch Pointing Control - The operation of the AFTI/F-16's flaperons was modified so they could operate as direct lift devices to allow new maneuvers: vertical translation at a constant attitude and pitch pointing at a constant altitude. These capabilities were called "decoupled modes", because the normally-connected pitch attitude and vertical motion now could be controlled independently. Thus, the nose could be raised to acquire a target without causing the aircraft to gain altitude, or precise flight path control could be maintained while holding the aircraft at a constant pitch attitude. These two capabilities could also be merged to produce "maneuver enhancement". For example, to enhance a pitch up maneuver, the flaperons were deflected automatically for a few seconds as the pilot pulled back on the stick. This added lift got the AFTI/F-16 moving vertically very quickly, even before the nose has risen to the desired climb attitude.

- Direct Sideforce and Yaw Pointing Control - The decoupling of the lateral and yaw motions allowed nose pointing and lateral motions that were independent of each other. Just like the vertical stabilizer and rudder, the canards produced a side force and a yaw moment when deflected. When the canards and rudder were used together and deflected in the same direction, the yaw moments canceled, resulting in a net side motion while the nose remained pointed straight ahead. With the two surfaces deflected opposite to each other, the nose could be pointed from side to side without any side motion.

- Digital Flight Control - The standard analog flight control computer was replaced with a digital computer. The complex maneuvers that the AFTI/F-16 would demonstrate required a continuous blending of all the control surface motions. While a maneuver may have appeared simple, it actually may have required the use of many control surfaces, each moving minute amounts at precise times. Making all these control motions were beyond the capability of the pilot. With the AFTI/F-16, the pilot only needed to move the stick in the direction he wanted the aircraft to move, and the computer then determined how to move each control surface to produce the desired motions. The complex mathematical equations needed to define these motions would have been extremely complex to implement on an analog based flight control system (the state-of-the-art in the 1970s), yet were relatively simple to program on a digital computer. Changes to these equations also were easier to make on a digital flight control computer. This was because the old program could be erased and replaced with a new one, whereas an analog flight control system required new circuit cards to be manufactured for each change. This technology promised greater flexibility and greater capability while reducing costs and complexity.

- Advanced Cockpit - The future battle environment likely will have more threats in a greater variety than ever before. At the same time, there likely will be fewer friendly aircraft to defeat these threats. Thus, more will be required of each aircraft and each pilot. Several new technologies already tested extensively in the laboratory held great promise. Their intent was to ease the pilot's workload while allowing him to perform a wider variety of tasks. The AFTI/F-16 incorporated an advanced cockpit that included dual cathode ray tubes, called multi function displays. Resembling small television monitors, they replaced many of the standard mechanical instruments. Pictures of the instruments, such as attitude indicator and flight director, were drawn on the displays. On different missions, such as cross country navigation, air-to-air combat, and air-to-ground combat, the pilot could call up different variations of these displays. In addition, the status of various systems that were monitored continuously by the computer could be displayed as needed, freeing the pilot from having to monitor stacks of dials. All this at the touch of a button.

- Voice Command System - Imagine telling the aircraft what it was to do, rather than having to push buttons or turn dials! The AFTI/F-16 included the first-ever flight test of a voice-actuated command system. The pilot could state a word or phrase and the action would occur. Of course there was a limited number of things the pilot could ask for, such as switching radios or arming various weapons, but the system would allow the pilot to keep his hands on the stick and throttle and his eyes outside the cockpit.

- Integrated Fire/Flight Control - The fire control system on an aircraft controls the target acquisition and weapon firing, while the flight control system makes the aircraft maneuver according to the pilot's commands. The merging of these systems, or integration, allowed a reduction in pilot workload never before achievable. In the past, to acquire a target, the pilot had to maneuver the aircraft until the target was near the center of his gun sight and manually position the radar to lock the fire control system on the target. On the AFTI/F-16, the gun sight was projected on the pilot's visor and the radar was slaved to the pilot's head position. To acquire the target, the pilot turned his head, put the gun sight on the target, then pressed a "designate" button on the control stick. With the target designated to the fire control system, the computer generated display gave the pilot continuous direction to the target. This left the pilot free to scan for other potential threats.

The AFTI/F-16 was completed and delivered to the Air Force in July 1982 and then ferried to NASA's Dryden Flight Research Facility at Edwards Air Force Base in California. It was operated throughout the early 1980s by a joint Air Force, NASA, Navy, and General Dynamics test team. The new systems were tested, refined,

ONE-OF-A-KIND RESEARCH AIRCRAFT

The AFTI/F-16 in formation with a T-38 chase aircraft. (USAF Photo)

and tested again. Typical to flight test programs, some things worked as expected, others had to be refined further. During this first phase of flight testing, over 100 test flights were made during 1982 and 1983. Ultimately, the AFTI/F-16 would fly three major flight test efforts during its career.

By 1984, more new technologies for potential future aircraft had been developed in the laboratory and were ready to be demonstrated in the AFTI/F-16. These all focused on making future aircraft more survivable during close air support. Future aircraft would have to stay low, hugging the terrain, to avoid detection. Thus it would be necessary to survey the target area, acquire and destroy multiple targets in a single pass, then depart. In the future electronic battlefield, loitering in the combat area and making a second pass at the target would both be too dangerous. In addition, missions would need to be flown day or night and in almost any weather. Some of the close air support technologies incorporated in the mid 1980s included:

•Internetted Attack - In the past, the pilot had to use his own eyes or on-board sensors to acquire and attack a target. He was often aided verbally by pilots in other aircraft, ground observer, or sensor operator in a nearby electronic warfare aircraft. With the internetted attack system, other sources of information, such as another attack aircraft, a surveillance drone, or even an orbiting satellite could transmit target data to the AFTI/F-16 to help it locate and lock onto a target automatically.

• Automated Maneuvering Attack System (AMAS) - This system allowed the AFTI/F-16 to make maneuvering attacks. In the past, the pilot had to align his aircraft with the target, then be at the proper airspeed, descent angle, and altitude before releasing his weapons. While aligning on the target, the attacker was very vulnerable. The AMAS system allowed maneuvering attacks in which the aircraft was never stabilized. It included an infrared tracker mounted in the starboard wing root.

• Enhanced Night Attack Helmet Mounted Display - At low altitude, at night, or in poor visibility, it is especially important for the pilot to keep his eyes focused outside the cockpit. Forward Looking Infrared Radar (FLIR) is a system that can see at night or in haze, delivering a visual picture of the terrain ahead. In the AFTI/F-16, FLIR signals were digitally enhanced to give the pilot a better visual picture of the surrounding terrain. The picture was displayed to the pilot on his helmet visor or head-up display, allowing him to keep his eyes outside the cockpit. In addition, any display that could be generated on the multi-function display in the cockpit, including systems information, could be projected on the pilot's visor.

The second phase of the AFTI/F-16 program began with a contract let to General Dynamics to modify the aircraft with these capabilities. Flight testing was performed from April 1985 until March 1987 and included over 150 test flights.

The third and final phase of the AFTI/F-16 program began in the late 1980s. The following major new system was added:

• Automatic Ground Collision Avoidance System (AutoGCAS) - One of the greatest dangers of close air support is the possi-

PART II: TESTBEDS

The AFTI/F-16 was also modified into a Close Air Support (CAS) configuration. (USAF Photo)

bility of hitting the ground. This can occur primarily in two different ways, either the pilot blacking out from pulling too many g's, or by flying too close to a terrain feature so that the pilot becomes "boxed in" and cannot avoid hitting it. Needless to say, the more rugged the terrain, the greater the possibility. The AFTI AutoGCAS system stored a digitized terrain map and used inertial navigation and a radar altimeter to provide a precise position and altitude on the digital map. When the aircraft was in danger of hitting the ground or flying into a mountain, the system provided a verbal and visual signal to the pilot. If the pilot did not respond quickly, either because he was preoccupied or had blacked out, the system would automatically level the wings and initiate a pull up to get the aircraft out of danger before it was too late.

The aircraft was modified for the third phase of flight testing, again by General Dynamics. Testing aircraft has always been expensive. With the AFTI/F-16 being the last pre-production model still flying, operating costs had become extremely high since many parts that were not quite like those on production F-16s had to be custom made. Other age-related problems were also appearing, such as numerous fuel leaks. As part of the aircraft modification for third phase flight testing, the wings, horizontal tail, and engine inlet were replaced with standard production items, making the aircraft much more supportable.

Test flights were completed in May of 1994, culminating with bomb drops and a live missile firing. In its final demonstration, the AFTI/F-16 and two F-18s flew to a mock target in the California desert. The F-18s attacked first. Subsequently, reconnaissance data obtained from other aircraft and satellites was precessed in real time, and the final target was changed in the onboard mission computer, all while the AFTI/F-16 was on its way to the target.

Despite talk of retirement, the AFTI/F-16 keeps on going. As of mid-1994, another phase of AFTI/F-16 flight testing was being prepared, this time to develop and demonstrate the technology needed to attack a target in bad weather, not just to navigate around or through the weather! The AFTI/F-16 may be around for years to come.

ONE-OF-A-KIND RESEARCH AIRCRAFT

YF-16 CCV/FLOTRACK Programs Testbed

AIRCRAFT TYPE: YF-16 (F-16 Prototype)
MISSION: Proving the concept of a digital fly-by-wire flight control system and use in FLOTRAK test program
OPERATING PERIOD: 1976 to 1977 (CCV), early 1980s (FLOTRAK)
RESPONSIBLE AGENCY: Air Force Flight Dynamics Laboratory
CONTRACTOR: General Dynamics

THE STORY
The concept of fly-by-wire flight control systems continued to interest the Air Force in the mid- to late 1970s, and the program continued with the so-called F-16 CCV (for Control Configured Vehicle) program. The highly successful program demonstrated that it would be possible to enhance the operational capabilities of future high-performance fighters by designing their control systems for non-conventional flight modes.

The F-16 that was used in this program was interesting in the fact that it was the first Fighting Falcon, the YF-16 prototype. The changes made to achieve the CCV configuration included attachment a pair of under-fuselage inlet duct canards – not unlike the previous F-4 Fly-By-Wire test aircraft – and mounted 30 degrees to the vertical.

Also, there was an auxiliary flight control computer which augmented the original control system. The CCV control panel selected the various control configured operating modes, activation switches on the stick, and special control configured mode displays on the instrument panel.

There were also changes made to the fuel system with the addition of three transfer pumps. Hydraulic actuators were used to cant the canards at different angles during flight with a range of plus-or-minus 25 degrees from their streamline position. In certain control modes, the canards actually operated in conjunction with the rudder to produce side force.

Program manager Dick Swortzel of the Air Force Flight Dynamics Laboratory explained that the CCV program was a excellent tool for designing new aircraft. He continued that the F-16 was an excellent aircraft for this type of testing, but that the non-conventional maneuvering capabilities of the aircraft were restricted by conventional control surfaces, such as the large vertical tail.

Swortzel continued that if a CCV fighter aircraft were to be designed from scratch, it would look quite different than this configuration. "With such an aircraft, the vertical tail would be much smaller in area to allow more side force authority, or the tail would

The F-16 Control Configured Vehicle (CCV) used the YF-16 prototype as the testbed aircraft. (USAF Photo)

PART II: TESTBEDS

The mounting location for the engine-mounted canards is shown here. The program demonstrated advanced maneuver capabilities applicable to future fighters. (USAF Photo)

be designed to swivel and function in a CCV mode as well as in a conventional mode."

The CCV F-16 demonstrated a number of lateral maneuvers which could be accomplished in wing-level flight. The capability was deemed to be effective for ground target tracking and air-to-air combat. The CCV capabilities were also assessed to be effective in tracking enemy vehicles on winding roads.

Another amazing capability the F-16 CCV demonstrated was the ability to point the nose up or down without power changes, or left or right while maintaining a constant track across the ground. These capabilities also made the aircraft less susceptible to air turbulence and wind gusts.

There were actually four phases to the CCV flight testing: functional checkout of the new configuration, evaluation of the multiple flight modes, CCV applications to operational situations, and relaxed stability testing which would be accomplished with the canard removed.

The forward canards for the CCV testbed are shown during the fabrication process. (USAF Photo)

One of the interesting effects felt by the CCV test pilots was the reaction to the .9Gs of sideforce on the aircraft. The pilots reported that the sensation was very fatiguing because there wasn't any support for the upper body. It was again determined that a CCV plane built from scratch would certainly have to address that phenomena, and determine ways to restrain the pilot during high lateral-G maneuvers. In all, some 85 test flights were conducted dur-

The F-16 prototype was the testbed aircraft for the CCV program, and later the FLOTRAK investigations. (USAF Photo)

ONE-OF-A-KIND RESEARCH AIRCRAFT

Here, the F-16 CCV testbed is employed in its second test program, the FLOTRAK program. The program attempted to develop a system to keep heavily-loaded fighters from sinking into the mud. (USAF Photo)

ing the program. It should be noted, though, that the program was strictly research in nature with no planning for CCV product improvement on production F-16s.

The F-16 CCV program, though, was just another step along the way toward advanced control systems. Like the previous F-4 Fly-by-wire program, contributions from this program would be applied to the more extensively altered, and long term program, the AFTI F-16.

FLOTRAK Program
Following its use in the CCV Program, this particularl YF-16 would later (early 1980s time period) be employed in the so-called FLOTRAK Program.

The purpose of this program was to devise a method of preventing heavily-loaded fighters from sinking into the ground, immobilized by their own heavy footprints. The solution which was undertaken in the FLOTRAK program, again managed by the Flight Dynamics Laboratory, was having the tires wrapped with FLOTRAK, a jointed, hard plastic shell that would lighten the aircraft's landing gear imprint by as much as two-thirds.

Support personnel would simply place the flattened track in front of the aircraft's tires – much as chocks are thrown down when aircraft are parked – and taxi the aircraft onto the FLOTRAK. The track-like devices wrapped around the tire and locked into place. Once the aircraft reached its take-off location, the FLOTRAK devices could be removed easily and was ready for immediate re-use on similar aircraft tires.

Testing was accomplished on the unique system at both Wright Patterson and Edwards Air Force Bases.

The concept showed that regardless of aircraft type, the FLOTRAK system reduced a landing gear's over ground pressures significantly. Without FLOTRAK on the F-16 main landing gear, for example, the tire over-ground pressure was 275 pounds per square inch. With FLOTRAK attached, the tire over-ground pressures dropped to about 80 pounds per square inch. With the nose landing gear, the pressures decreased from the normal 215psi to about 80psi.

Each FLOTRAK devise for the F-16 main landing gear weighed about 97 pounds, while the nose unit was only 27 pounds. The FLOTRAK units were fabricated of a polyester elastomeric material that was high-temperature resistant.

PART II: TESTBEDS

F/A-18 EPAD Program Testbed

AIRCRAFT TYPE: F/A-18
MISSION: Test of an electrically-powered flight control design
OPERATING PERIOD: Mid-1990s
RESPONSIBLE AGENCY: U.S. Air Force/NASA/Navy
CONTRACTOR: Lear-Seigler

THE STORY

It might have seemed like fiction just a few years back, but the concept of electrically-powered components leading to an all-electric aircraft could occur in the not-to-distant future.

The effort was called the Electrically Powered Actuation Design (EPAD) validation program, and was a joint USAF, Navy, and NASA venture with the Aeronautical Systems Center (ASC) at Wright Patterson Air Force Base having overall responsibility with the F/A-18 being the test aircraft.

The F/A-18 was selected for this program because it was a totally fly-by-wire aircraft, making it fairly easy to modify for a single actuator. More importantly, the performance requirements were significant step-ups in both response rates and power density – the actual keys to electric actuation on fighter aircraft.

The initial flight testing for the EPAD program took place at NASA Dryden at Edwards Air Force Base, California in the May-July 1993 time period. The second phase of the testing started in the fall of the same year.

Captain Chris Hansen, the EPAD Program Manager, explained, "the EPAD would validate electrically-powered flight control actuators. If successful, electric actuators would replace conventional [hydraulic] powered actuators for control surfaces such as flaps and ailerons, and eventually eliminate the need for a central hydraulic system on an aircraft."

This innovation was assessed to greatly increase the reliability of an aircraft flight control system, as well as easing maintenance. Electric actuators are self-contained (not connected to the central hydraulic system) and line-replaceable units. As such, when an electric actuator fails, it is much easier to replace due to simple mechanical and electrical connections, eliminating the need for hydraulic specialists.

A major advantages of such a system is that it is environmentally friendly, with no hydraulic fluid to spill, dispose of, or store. Also, the future of this concept is not limited to aircraft. Literally every piece of equipment that used hydraulics could benefit from this technology.

The three-phased EPAD flight test program evaluated three types of electrical actuators. The flight test program consisted of about two dozen missions per actuator, each mission lasting approximately one hour.

In the initial tests, a "smart actuator" replaced the conventional actuator powering the aileron on the left wing of the F/A-18 test aircraft.

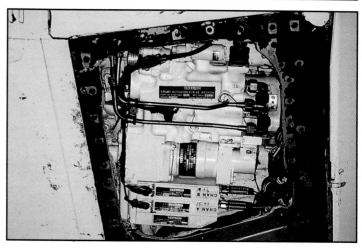

The so-called EPAD program used an F/A-18 testbed aircraft to test an electrically-powered activation system. This photo shows the installation details. (USAF Photo)

The "smart actuator" was a conventionally-powered device using the central hydraulic systems, but the control electronics were mounted on the actuator itself. The device provided self-contained, fail-operate/fail-safe operation, as does the normal F/A-18 aileron actuator. The unit sensed the input from the flight control computer over a redundant optical bus and electrically told the actuator to move. The actuator moved the aileron, though, by hydraulic power.

The F/A-18's flight control computer was not altered to accommodate the electrical actuator, but interface boxes translated the flight messages into the language that the electrically controlled actuator could understand. Another unique aspect about the interface box was that it used fiber optics to connect with the actuator, one of very few fly-by-light technologies used in a testbed aircraft anywhere in the world.

The idea, though, for electrical actuators is not new the idea being investigated previously in the laboratory. Then in 1985, an electro-mechanical actuator was tested on the left aileron of a C-141 transport aircraft attached to the 4950th Test Wing at Wright Patterson AFB, Ohio. However, the response rates were not sufficient to justify for the goal of placing electric actuators on a fighter aircraft.

It should be mentioned as a final note, that even though the Air Force funded a significant portion of the research program, the involved contractors also provided significant funding because of the promise of the concept.

Besides the prime Lear Seigler contractor, there was also involvement from Martin Marietta, MPC Products, and HR Textron.

ONE-OF-A-KIND RESEARCH AIRCRAFT

F/A-18 HARV Program Testbed

AIRCRAFT TYPE: F/A-18
MISSION: To demonstrate high angle-of-attack flight
TIME PERIOD: 1980s and 1990s
RESPONSIBLE AGENCY/S: NASA/Ames/Langley/Lewis
CONTRACTOR: McDonnell Aircraft Company

THE STORY

Angle of Attack (AOA) is the nomenclature used to describe the angle between an aircraft's body and wings relative to its actual flight path. During maneuvers, having a capability at high angles-of-attack; when the nose it kicked up while the aircraft continues in its original direction; can provide significant advantages in most combat situations.

Flying at this attitude, problems are created when the air-flow around the aircraft becomes separated from the airfoils. At high angles-of-attack, the forces produced by the aerodynamic surfaces – including lift provided by the wings – are greatly reduced. This flight attitude often results in insufficient lift to maintain altitude or control of the aircraft.

The NASA HARV program produced technical data at high angles-of-attack to validate computer analysis and wind tunnel research. Successful validation of this data enabled engineers and designers to understand better the effectiveness of flight controls and airflow phenomena at high angles-of-attack. The goal of this ambitious program, simply stated, was to provide better maneuverability in future high performance aircraft and make them safer to fly.

The HARV program used a modified version of a pre-production F/A-18 Hornet. The testbed retained its standard GE F404 powerplants, each capable of 16,000 pounds of thrust in afterburner. Typical takeoff weight for the HARV is about 39,000 pounds with 10,000 pounds of internal fuel, the thrust vectoring paddles, and a unique spin recovery parachute system installed.

Several of these pre-production aircraft were given to NASA by the Navy in the 1980s to replace the F-104s previously used for a variety of test missions.

Even without the high angle-of-attack enhancement modifications, the F/A-18 fighter was already an excellent high angle-of-attack performer. In its standard configuration, the plane had no restrictions at normal center-of-gravity positions. It was this characteristic that made the model an excellent choice to accomplish this particular test program.

The program began in 1987 with the first of three phases. The initial phase lasted for two and one-half years and consisted of 101 flights with a specially-modified F/A-18 HARV (High Angle of Attack Research Vehicle) which flew up to angles-of-attack approaching 55 degrees. To measure the test plane's performance, visual studies of the airflow over various parts of the aircraft were performed.

Special smoke was released through small ports just forward of the F/A-18's leading edge extensions near the nose, and photographed as it followed airflow patterns around the aircraft. Also photographed in the airflow were small pieces of yarn taped on the aircraft, along with an oil-based dye released onto the aircraft surfaces from 500 small orifices around the aircraft nose. Additional data obtained included air pressures recorded by sensors located in a 360-degree pattern around the nose and at other locations on the aircraft. All these measurements and flow visualization techniques helped engineers to correlate theoretical predictions with what was actually happening.

Phase two of the HARV program began in the summer of 1991, using a considerably-modified HARV vehicle for new test mission requirements.

Three spoon-shaped, paddle-like vanes were mounted on the airframe around each engine's exhaust. The devices deflected the jet exhaust to provide both pitch and yaw forces to enhance maneuverability at high angles-of-attack when the aerodynamic controls were unusable or less effective than desired. Also, the engines had their external exhaust nozzles removed to shorten the distance by

This F/A-18 pre-production fighter aircraft was modified for the HARV Program. The program started in 1987 and was continuing into the 1990s. Clearly visible is the paddle actuation system for deflecting engine exhaust. (NASA Photo)

PART II: TESTBEDS

The capability to maneuver at high angles-of-attack received high attention in the 1990s with the HARV program which was a multi-center NASA program. (NASA Photo)

two feet the vanes had to be cantilevered. The unique thrust-vectoring modification precluded the HARV aircraft from being able to accomplish supersonic flight. The system added over two thousand pounds to the craft's weight.

The modifications enabled the HARV to be much more effective at higher angles-of-attack, up to near 70 degrees. Since the airraft could hold the high AOA for longer time periods, much more data was available for collection.

The envelope expansion flights were completed in February 1992. Demonstrated capabilities included stable flight at 70 degrees, up 15 degrees from the previous high of 55 degrees and rolling at high rates at 65 degrees angle of attack. This phase of the program was concluded in late 1993.

During early 1994, the final phase of this program was undertaken with additional modifications of the HARV test aircraft being accomplished. Changes included the addition of movable strakes which were mounted on both sides of the nose to provide enhanced yaw control at high angles-of-attack. The four-foot-long, six-inch-wide strakes, are hinged on one side and mounted flush to the forward sides of the fuselage.

At low angles-of-attack, the strakes were folded flush against the aircraft skin, while they were extended at higher angles in order to interact with vortices generated along the nose. As a result, large side forces for enhanced control were produced. Early research indicated that the strakes could be as effective as rudders at the lower angles of attack.

A majority of the high angle-of-attack modifications were done on the rear portion of the F/A-18's fuselage as is clearly evident in this photo. (NASA Photo)

ONE-OF-A-KIND RESEARCH AIRCRAFT

F/A-18 Systems Research Aircraft Testbed

AIRCRAFT TYPE: F/A-18
MISSION: Testbed for advanced control systems
OPERATING PERIOD: 1993 to mid-1990s
RESPONSIBLE AGENCIES: NASA Dryden Flight Research Center, NASA Lewis Flight Research Center, US Navy
CONTRACTOR: McDonnell Douglas

THE STORY

The F/A-18 SRA (Systems Research Aircraft) has been involved in a number of NASA test programs, but its most important mission has been a series of tests that could result in lighter, more fuel-efficient aircraft concepts with more capable control and monitoring systems using a so-called Fly-by-Light concept.

The initial flight of the SRA testbed aircraft demonstrated successful operation of several fiber optic sensors including rudder position, rudder pedal position, leading-edge flap position, nose wheel steering, total pressure, and air data temperature.

This effort was instituted in 1993 with the Fiber-Optic Control System Integration (FOCSI) program. The purpose of the program was to examine fiber-optic engine controls and set the stage whereby a future transport aircraft could be fitted with partial fiber-optic or fly-by-light flight and engine control systems. Another part of the FOCSI program was to examine passive optical sensors and fiber-optic data links.

The FOCSI tests using the SRA plane had a goal of developing fiber optic systems – small bundles of light-transmitting cable – that weigh less and take up less space than the copper wiring in today's aircraft.

Darlene Mosser-Kerner, SRA Chief Engineer, explained, "Replacing copper wires, where appropriate, offers an unsurpassed ability to transmit commands and data between parts of an airplane. The FOCSI flight tests will help develop fiber optic components that will carry signals to and from flight controls in tomorrow's civilian aircraft."

Weight and fuel savings in transport-type aircraft engineered with fiber optic control systems could be substantial compared to designs with traditional copper wiring. The long copper cables not only weigh more than fiber optics, but also must be shielded. Shielding is essential to keep the wires from accidently broadcasting electrical signals that might cause interference.

The fly-by-light system promised a number of other advantages including (1) Reduction of signal transmission, (2) Cut aircraft certification costs, and (3) Boost aircraft first safety be eliminating possible ignition sources.

The F/A-18 test bed aircraft was heavily instrumented, including sensors in (1) the left trailing and right leading edges, (2) the rudder and left stabilizer, and (3) on the engine power lever control. In addition, sensors were also added to determine aircraft pitch stick, rudder pedal and nose-wheel positions.

Another phase of the program involved the fitting of a General Electric F404 powerplant with a number of optical sensors. Sensors which monitored fan speed, inlet temperature, core speed, compressor inlet temperature, and turbine exhaust temperature. The optical engine control portion of the FOCSI program is expected to begin in the mid-1990s time period.

The lessons learned on the FOCSI program will be applied to the development of fiber optic sensors used to actively control a rudder actuator and flight critical stabilator actuator in late FY 1995.

This NASA F/A-18, specially modified to test the newest and most advanced system technologies, on its first flight in May 1993, at the Dryden Flight Research Center, Edwards, California (NASA Photo)

PART II: TESTBEDS

The F/A-18 SRA aircraft evaluates technologies that will benefit both civilian and military aircraft. (NASA Photo)

The FOSCI program is a part of the Fly-By-Wire/Power-By-Wire (FBW/PBW) program which itself is part of the NASA Advanced Subsonic Technology Initiative.

A later program, called the More Electric Aircraft Program, also using the SRA testbed, was an Air Force initiative to develop aircraft control systems in which electric motors would be used instead of hydraulic systems to move control surfaces.

"The benefits of this concept include 30-to-50 percent reduction in ground support equipment, five-to-nine percent reduction in fuel consumption, and 600-to-1000 pounds reduction in take-off weight for commercial airliners," explained Joel Sitz, Chief Engineer for the program.

With this and other scheduled research programs planned, the SRA should remain a busy and productive testbed aircraft throughout the 1990s.

Other that the external markings, there is nothing to identify this F/A-18 from its operational brothers. (NASA Photo)

ONE-OF-A-KIND RESEARCH AIRCRAFT

JF-100 Variable Stability Testbed

AIRCRAFT TYPE: North American F-100C
MISSION: In-Flight Simulation/Flight Control Research
OPERATING PERIOD: 1960-1964
RESPONSIBLE AGENCY: NASA Dryden Flight Research Center

THE STORY

NASA's variable stability JF-100 is another example of a research aircraft for which very little information is centrally located and readily available. Over the years, as memories of the aircraft faded, all that remains were scattered references to it and bits of information in other reports and documents.

The JF-100 was built from an Air Force F-100C by NASA's Ames Research Center, and transferred to NASA's Dryden Flight Research Center in 1960. The "J" designation refers to it being modified for special test missions, but not so extensively that it could not be returned to being a standard F-100. The aircraft obviously was acquired from the Air Force and carried the registration number 53-1709, but no information about its earlier career was available.

The aircraft incorporated an analog variable stability system that modified the pitch, roll, and yaw motions. In addition, the ailerons were driven independently by the flight control computer, and could be deflected in the same direction to produce direct lift. The flight control computer could completely control the ailerons and rudder, but the horizontal stabilizer was limited to five degrees of travel about the trim point. This was because the F-100's all-flying tail was extremely powerful and the aircraft could loose control quickly if a computer-generated command was calculated improperly.

Test flights were recorded using two 16-mm motion picture cameras, and a photographic oscillograph (basically a strip chart recorder that wrote with a light beam on photo-sensitive paper) which could record 26 channels of aircraft data.

It is interesting to note that the JF-100 was a single-place aircraft. The pilot performed the functions of both the evaluation and safety pilots! In the event a configuration proved difficult to control, the pilot disengaged the computer by pressing one of two switches to disengage the computer, again making the aircraft fly like an F-100. There were also numerous automatic safety monitors that would kick off the fly-by-wire system if certain pre-set values were exceeded.

The JF-100 performed several research programs during its brief career. These included early studies of supersonic transport flying qualities, developing special piloting techniques for the X-15 (quick, discrete pulses to the control stick, then back off and wait for the aircraft to respond), and investigating the use of direct lift to improve precision during in-flight refueling.

The JF-100 was removed from service as a variable stability aircraft at NASA Dryden in 1964, but its final disposition could not be determined. The information is most likely buried in various reports somewhere, waiting to be rediscovered by a future researcher.

NASA modified an F-100 Super Sabre to the JF-100 during the 1960s to test a number of flight regimes. It's shown here on the lake bed at Edwards Air Force base. (NASA Photo)

PART II: TESTBEDS

F-102 Low L/D Testbed

TYPE OF AIRCRAFT: F-102A
MISSION: Testing of low Lift/Drag concepts
TIME PERIOD: 1960s
RESPONSIBLE AGENCY: NASA Dryden Flight Research Center

THE STORY
This F-102A Delta Dagger was one of the first, if not the very first, testbed aircraft to investigate low lift/drag (L/D) performance. A number of other higher-performance aircraft, like the follow-on F-104 testbed would continue the testing, but this testbed set the initial standards.

In order to accomplish the required testing, the aircraft was modified with a larger speed brake for the following programs:

X-20 DynaSoar Program
Like the F5D Skylancer testbed, the F-102 was also used in landing-approach simulations of the X-20 DynaSoar space glider in the early 1960s. Circular and straight-in approach and landing patterns were examined and the same positive conclusion was reached.

Pilots thought that the circular patterns allowed more precise positioning at the end of the runway than did a straight-in approach. The L/D ratios examined during the program were in the 2-4 range, with this testbed being extremely effective at the lower numbers. This capability was achieved because of the F-102's extremely low wing loading.

X-15 Program
Like a number of other NASA testbed aircraft, this F-102 was also used in the landing and approach phase simulations for the rocket-powered spacecraft. The F-102 testbed was used to establish the X-15 landing pattern and to train the pilots in the proper landing procedures.

One of the first investigations into the benefits of low L/D flight was accomplished by this F-102 delta-wing fighter. (NASA Photo)

ONE-OF-A-KIND RESEARCH AIRCRAFT

NASA F-104 Testbeds

TYPE OF AIRCRAFT: Various models of the F-104 Starfighter
MISSION: Various missions
TIME PERIOD: 1950s to 1990s
RESPONSIBLE AGENCY: NASA Dryden Flight Research Center
CONTRACTOR: Lockheed

THE STORY
In 1994, the F-104 fleet of testbed aircraft at the NASA Dryden facility celebrated their 37-year anniversary with a total of over 18,000 flights to their credit. The first NASA (then NACA) flight of an F-104, tail number 961 (later renumbered to 818), marked the beginning of a long and productive series of research and support flights. Through the years, there have been just about every imaginable mission flown by the versatile Starfighter testbeds.

With eleven different F-104s in service at different times, Dryden used the model for basic high speed and bio-medical research as well as airborne simulation of the X-15 and lifting body programs. Aircraft number 820 set a record of over 100 flights in a one-year period. Details of some of the more significant programs are:

Low L/D Ratio Approach and Landing Testing
In the late 1950s, an F-104A was used to simulate low L/D approach and landing techniques. By suitably scheduling thrust and drag-producing devices, a maximum L/D of 2.8 and a wing loading of 75 lb/square feet were obtained.

X-15 Approach and Landing
An F-104A testbed was also used to simulate the landing and approach phase of the X-15 rocketship. The X-15 was a low L/D vehicle that could only be landed dead stick at fairly high speeds, so it was important to establish the landing pattern and to train the pilots in the proper procedure.

Several other F-104As were used to evaluate circular and straight-in approach procedures under simulated X-15 mission conditions. Experience with the Starfighters indicated that an L/D of approximately 2.5 represented a practical lower limit where the vehicle could be landed safely.

Approach and Landing Visibility Testing
The F-104s were used in the 1960s to investigate visibility requirements for approach and landing for low L/D aircraft.

The first effort, using an F-104B, with an indirect viewing system that had two wide-angle overlapping periscopes with stereoscopic vision. The periscopes were mounted on the canopy between

This pair of NASA F-104 testbed aircraft long performed low L/D flight testing. (NASA Photo)

PART II: TESTBEDS

You better believe that NASA loved the F-104 as a testbed aircraft. Here the fleet is aloft in formation. (NASA Photo)

Now retired, the Number 826 F-104 was used to test thermal insulation tiles during its final missions. (NASA Photo)

the front and rear cockpits, and the image was shown to the evaluation pilot in the rear seat.

In another visibility study, additional test equipment was installed including a radar altimeter. An attempt to use an early head-up display (HUD) was unsuccessful in that the pilots found the information unreadable.

The third F-104 limited visibility study, also flown in the 1960s, involved masking the forward view so the pilot had to relay on the field of view from side windows to land. The results showed that a large amount of the forward field of view could be obscured before the landing performance suffered to any extent.

The TF-104G testbed Starfighter was involved with testing for the National Aerospace Plane (NASP) program. A number of windows, selected to match those proposed for the NASP were examined using straight-in approaches. Later investigations using that aircraft saw the installation of a folded-mirror optical viewing system, also for NASP applications, with the investigations centering on low L/D approaches and precision landings.

Reaction Control System Testing

An instrumented YF-104A had a reaction control system installed and tested in the 1960s. The set-up was installed to obtain flight experience with jet reaction controls at low dynamic pressures. The testing was again accomplished to determine the landing qualifies of the aircraft at the low dynamic pressures encountered in the X-15 program.

PART II: TESTBEDS

NF-104 Aerospace Trainer

AIRCRAFT TYPE: Lockheed F-104A
MISSION: Manned Spacecraft Transition Training
OPERATING PERIOD: 1963-1971
RESPONSIBLE AGENCY: U.S. Air Force Aerospace Research Pilot School
CONTRACTOR: Lockheed-California Company

THE STORY

In the early 1960s, the United States was beginning to make its presence in space a national priority. Project Mercury was showing than man could survive in space and perform essential functions. The X-15, the first of a planned series of winged spacecraft, showed promise that future operational spacecraft could be flown into space and then flown to conventional landings. The NF-104 Aerospace Trainer was used to train the test pilots who would fly this new generation of aircraft into orbit by allowing them to zoom to 100,000+ feet in a full pressure suit, experience zero "g", and use reaction control to control their aircraft.

The NF-104s were modified Lockheed F-104A Starfighter aircraft. Three were built by the Lockheed-California Company. Major modifications included the addition of a 6,000-pound thrust rocket engine at the base of the vertical tail, reaction control thrusters in the nose and in each wing tip, a larger vertical tail, increased wing span, tanks to store the rocket propellants, provision for a full pressure suit, a cockpit hand controller to operate the reaction control thrusters, and modified cockpit instrumentation. Combat equipment that was no longer needed, such as the gun, fire control system, tactical electronics, and auxiliary fuel tanks were removed.

The NF-104s were operated by the Air Force's Aerospace Research Pilot School at Edwards Air Force Base, California, since renamed the U.S. Air Force Test Pilot School. The idea for the NF-104s was generally credited to astronaut Frank Borman, who was a student at the school in 1960 and an instructor in 1961.

A contract was awarded to the Lockheed-California Co. for modifying three aircraft, and the first flight was completed in June 1963. Extensive flight testing was required, followed by development of the student curriculum flights. The aircraft did not become operational at the school until May 1968. Only a few student in each class earned the privilege of flying one of the NF-104s, for a total of only about 35 students.

The highest altitude achieved by any of the NF-104s was 121,800 feet by Maj. Robert Smith during acceptance testing. He used a 70 degree climb, but the climb angle was eventually restricted to 45 degrees when flown by student test pilots because of safety concerns.

Each student selected for the program performed two NF-104 flights. The typical flight syllabus started with taking off on jet power, climbing to 30-40,000 feet, and accelerating to 1.7-1.9 Mach. At this point the pilot ignited the rocket engine and pitched the nose up to start the steep climb. After about two minutes the aircraft passed through 80,000 feet, by which time the jet engine flamed out and the rocket engine ran out of fuel. The pilot now began a parabolic arc to his peak altitude. During the parabolic arc, the pilot experienced "weightlessness" for about one minute and used the side stick to fire the reaction control rockets to control the aircraft's pitch, roll, and yaw motions. Once at a lower altitude, the pilot restarted the jet engine and made a conventional landing. The entire mission lasted about 35 minutes from taxi to landing and was performed in a full pressure suit.

To prepare for their NF-104 flights, each student had to perform five preparation flights in standard F-104s. First was a pressure suit familiarization flight. The aircraft was flown to high altitude with the cockpit depressurized so the student could experience flying in a fully pressurized suit. The second flight was with an instructor in a two-place F-104 to practice the zoom profile. These were performed at a 30 degree climb angle reaching a peak altitude of 70-80,000 feet. The next three preparation flights were flown solo, repeating the climb maneuver, but increasing the climb angle to 45 degrees and reaching peak altitudes of 90,000 feet.

Two NF-104 flights almost ended in disasters. The first occurred on 10 December 1963, in which one of the aircraft was destroyed. Col. Chuck Yeager, then the Commandant of the Aerospace Research Pilot School, was making an altitude record attempt that was officially sanctioned by the Air Force. Yeager performed a 60 degree climb. One minute after reaching his peak altitude, while at 101,595 feet, the aircraft went into an uncontrollable yawing and rolling motion. Yeager was never able to recover and finally bailed out at 8,500 feet. When he separated from the ejection seat, it crossed in front of him and the rocket nozzle hit his face shield, breaking it. The combination of the red-hot nozzle and oxygen in his helmet caused a flame that burned the left side of his face and neck and set several parachute cords on fire. With his gloved hands, he beat out the flames, resulting in burns to his left hand. Yeager was hospitalized for two weeks following the crash.

The second incident occurred on 15 June 1971 on a student flight piloted by Capt. Howard Thompson. While at 35,000 feet and Mach 1.15, Thompson attempted to light the rocket engine, but instead heard a loud explosion. The chase plane reported that the rocket engine and half of the rudder were missing. Thompson flew back to Edwards AFB using the normal jet engine and made a safe landing.

The program was terminated because it was decided that the Air Force would no longer perform space flight training, the mission being acquired by NASA. The lone surviving NF-104 made its last flight in December 1971. It is currently on static display in front of the Air Force Test Pilot School at Edwards AFB, pointing 45 degrees nose-up, toward the sky.

The NF-104 research aircraft is now on display at the Air Force Test Pilot School at Edwards Air Force Base. (Steve Markman Photo)

ONE-OF-A-KIND RESEARCH AIRCRAFT

The NF-104 is shown in a zoom climb with the aid of its rocket engine. (USAF Photo)

PART II: TESTBEDS

F-100, F-106 Turbulence Research Testbeds

AIRCRAFT TYPES: F-100, F-106
MISSION: To examine the effects of turbulence on flight operations
TIME PERIOD: Late 1960s
RESPONSIBLE AGENCY: U.S. Air Force/4950th Test Wing

THE STORY

It had become a serious problem during the 1960s that the Air Force decided needed to be addressed. There had been a number of crashes that had been attributed to aircraft flying through turbulence, so an extensive test program was initiated by the 4950th Test Wing at Wright Patterson AFB using a number of highly-instrumented testbed aircraft.

F-106A Testbed Aircraft

Initial testing involved the use of a modified F-106A to examine the frequency and magnitude of low level gusts in the vicinity of mountains. Flying out of Kirkland Air Force Base, New Mexico, a highly-instrumented F-106A recorded its time, position, weather, and pilot conversations. During this period, there were some five dozen test flights logging a total of 89 hours.

The findings from this portion of the program showed that turbulence near the mountains was strong enough to destroy an aircraft, and needed to be taken into account in future aircraft designs.

F-100 Testbed Aircraft

Starting in 1967, the 4950th began studying the effects of medium altitude clear air turbulence, employing a highly-instrumented F-100. The test program was started out in the western United States in the mountain regions around Hill Air Force Base, then moving to the southeast United States, and then to the Griffis AFB, New York area.

In addition to studying clear air turbulence, the Air Force was also interested in turbulence associated with thunderstorms. With aircraft flying higher and faster, the Air Force wanted more information about these dangerous weather situations.

Rough Rider was the name given to a program for turbulence testing. The logo can be seen on the tail of the F-100 testbed aircraft. (USAF Photo)

A special probe, with sideslip and angle-of-attack sensors, was carried on the forward section of the F-100 turbulence test aircraft. (USAF Photo)

The National Severe Storm Laboratory (NSSL) teamed up with the Air Force for this effort using a unique F-100 called the ROUGH RIDER, which carried a bucking bronco as its logo, supposedly a tribute to what it was like flying through the storms.

This particular F-100 testbed aircraft was used during the 1960s in wind turbulence testing. (USAF Photo)

ONE-OF-A-KIND RESEARCH AIRCRAFT

F-111 TACT/AFTI Testbed

AIRCRAFT TYPE: General Dynamics F-111A
MISSION: Research for New Wing Designs
OPERATING PERIOD: 1973-1989
RESPONSIBLE AGENCY: NASA Dryden Flight Research Center, Air Force Flight Dynamics Laboratory
CONTRACTORS: General Dynamics, Boeing

THE STORY

TACT Program

Following the success of the supercritical wing research program performed by NASA Dryden on its F-8 testbed, the Air Force and NASA Dryden agreed to a joint program to test the military utility of this new airfoil shape. The program was named TACT, which stood for Transonic Aircraft Technology. An F-111A was chosen as the testbed aircraft, primarily because its variable-sweep wing would produce data applicable to a wider range of wing designs.

The Air Force provided a pre-production F-111, the thirteenth off the assembly line, serial number 63-9778, to NASA Dryden in February 1972. The first thing done was to instrument the aircraft and fly twenty four data flights to collect baseline data on performance and airflow. This data would be used later to determine improvements achieved with the supercritical wing. These flights were performed in the summer and fall of 1972, and gathered data out to Mach 2 and 50,000 feet altitude.

While these flights were being performed, General Dynamics built the new wings, using a similar supercritical airfoil shape as was tested during the F-8 program, but utilizing a wing planform very similar to a standard F-111. By early 1973, the TACT/F-111's wings were removed by NASA Dryden personnel. The new right wing was shipped to Dryden for installation, while the new left wing was shipped to Wright-Patterson Air Force Base for testing, then to Dryden.

Test flights of the completed TACT/F-111 aircraft were performed throughout the mid-1970s. They confirmed the test results gained earlier in the F-8 program, and showed that similar results could be expected for military applications. In addition, because the tests were performed at a variety of wing-sweep angles, designers now had a wealth of data that would help them to predict with great accuracy the performance of future wing designs using the supercritical airfoil shape.

In 1979 and 1980, the TACT/F-111 was used as a testbed to perform research on natural laminar flow. NASA had already performed other programs over the preceding twenty years and proved that laminar flow greatly improved wing efficiency and decreased drag. These past efforts produced laminar flow by sucking air into the wing to keep the flow smooth along the surface. Natural laminar flow, on the other hand, relies on a very precise airfoil shape rather than mechanical means to maintain the smooth flow.

Instead of constructing new wings, six-foot-wide "gloves" were fitted over each wing. These were made of fiberglass skin over an inner core of polyurethane foam and made in place right on the aircraft wing. The right glove was instrumented to measure pressure at various points. For this short series of test flights, the TACT program markings were removed from the fuselage. Although the test results were encouraging, the six-foot span was deemed too

After completion of the TACT Program, gloves were installed to investigate natural laminar flow. (NASA Photo)

PART II: TESTBEDS

The rear view of the F-111 AFTI testbed displays all trailing edge surfaces smoothly deflected. (NASA Photo)

narrow. At high wing-sweep angles, the airflow over the glove was contaminated by turbulent air spilling over from the inboard section of the wing.

AFTI Program
Toward the end of the TACT program, 63-9778 was selected for another joint Flight Dynamics Lab/NASA Dryden program, Advanced Fighter Technology Integration, or AFTI. Believe it or not, this was a program to make a modern version of an idea that preceded the Wright Brothers first aircraft. Early gliders did not have ailerons for roll control. Rather, they used cables to twist the wings. The idea was simple enough and allowed a greatly reduced number of mechanical components inside the wing. But by World War I, aircraft designs required the elimination of as much external bracing as possible, resulting in the wings becoming stiffer. The simple wing warping mechanism could no longer be used. Despite its simplicity and aerodynamic advantages of a smooth, continuous surface, the concept was relegated to the history books.

However, the wise use of history books allows ideas to be kept alive, even ideas that are not currently workable. Such was the case with wing warping. As aircraft flew faster, designers realized that major sources of drag were the hinges that held the ailerons and flaps to the wing, and the gaps between these surfaces. Indeed, anything that disturbed the airflow over the wing caused drag. Designers always knew of the advantages offered by a smooth wing that could bend to duplicate the functions of ailerons and flaps, but there was no way to make it work with an aluminum skin.

By the 1970s, enough advancements had been made with composite materials that it was felt a skin made of such material could offer the flexibility and strength needed, yet resist cracking as would a metal skin. Composites are materials made of either metallic or glass fibers held in place by an epoxy binder. By aligning the fibers in a particular arrangement, the strength characteristics of the finished composite can be tailored to meet very specific structural needs.

This view of the AFTI F-111 shows its leading edges smoothly deflected and the trailing edges slightly deflected. (NASA Photo)

ONE-OF-A-KIND RESEARCH AIRCRAFT

If you look closely on the wing, you can pick out the supercritical airfoil shape. (NASA Photo)

Boeing was contracted to design and build the variable camber wing and began work in early 1979. The airfoil shape was the same supercritical design used in the TACT program, and the same wingbox was retained. The leading edge flaps were replaced with a single-piece unit and the entire trailing edge was replaced with a three-section trailing edge flap. The two inboard sections functioned as flaps, while the outboard section functioned as the aileron. The skins were made from a fiberglass material. Despite the complex internal system to move the sections, overall, the wing had fewer parts than a standard F-111 wing.

The wing was given the name "Mission Adaptive Wing." The intent was not only to use the new control surfaces for maneuvering, but to change the camber throughout the mission so that the wing would always have the most efficient shape and provide maximum cruise efficiency. A conventional aircraft wing can only be most efficient at one altitude, airspeed, and weight. For any aircraft, the final wing design is a compromise, considering all aspects of the mission, such as take-off, cruise, combat maneuvering, and landing, each with many combinations of internal and external loads and at a variety of weights. With the Mission Adaptive Wing, the computer would reconfigure the wing constantly to be always at the optimum position, regardless of the mission, without any pilot intervention.

The AFTI/F-111 made its first flight from Dryden Flight Research Center in October 1985, reaching 15,000 feet and 300 knots. This was its first flight in over four and a half years. By July 1986, the AFTI/F-111 made its first supersonic flight. The first phase flight testing consisted of 26 flights, and collected data by manually setting the flap positions and determining cruise performance at a variety of speeds, altitudes, and weights. These flights continued through 1986.

The second phase involved computer control of the flaps, so that the wing shape would be constantly and automatically adjusted in order to keep an optimal shape. Four automatic modes were programmed into the computer. These were:

• Maneuver Camber Control - The computer automatically positioned the leading and trailing edge flaps to specific deflections depending on factors such as speed and angle-of-attack so that the aircraft would always be able to operate throughout its full maneuvering envelope.

• Cruise Camber Control - The computer constantly adjusted the leading and trailing edge flaps to produce the lowest drag. This was performed in an "adaptive mode", in which the flaps were constantly manipulated in minute amounts to see which setting produced the greatest speed. The simplicity of the concept was that it eliminated the need for complex formulas to determine the best flap settings, and would work for any aircraft configuration, at any weight, speed, or altitude.

PART II: TESTBEDS

- Maneuver Load Control - As an aircraft maneuvers, the lift produced causes the wings to bend upward. Quite often, the bending load reaches its limit long before the maximum g's are reached. The bending is produced by lift generated on outboard sections of the wing, which must be transferred inboard through the wing structure. To reduce bending, the AFTI/F-111's "ailerons" could both be deflected up during high g loads, so that little lift would be produced by the outboard sections. The lost lift was easily made up by the rest of the wing, but the bending tendency was greatly reduced. Thus, the aircraft could pull higher g's with no change to the wing structure because the bending load was greatly reduced.

- Maneuver Enhancement/Gust Alleviation - In this mode, the flaps were blended with the elevator at allow the aircraft to pitch up much more quickly than with elevator only. Also, sensors determined when the aircraft was being hit by wind gusts and automatically moved the control surfaces to counter the effects of the gust. This accomplished two things. First, it reduced pilot fatigue, especially on low level missions. Second, it made the aircraft a more stable platform for dropping unguided weapons.

AFTI/F-111 Phase II flights began in August 1987 and continued until October 1988. Fifty-nine flights were accomplished, totaling 143 hours. During these tests, all performance predictions were accomplished or exceeded, including the demonstration of a 25 percent range increase, a 25 percent sustained g increase, and an 18 percent peak g increase. The wing also required 35 percent fewer maintenance hours because all wing mechanisms were sealed inside the wing, away from dust, dirt, ice, and snow.

Its mission completed very successfully, the TACT/AFTI F-111 was retired at the completion of these programs. In February 1990, it was placed on static display at the flight test museum at Edwards Air Force Base.

ONE-OF-A-KIND RESEARCH AIRCRAFT

Air Force Transport Testbeds

AIRCRAFT TYPES: KC/C-135, C-141, C-130, and C-131
MISSION: To perform a wide variety of test missions
TIME PERIOD: 1950s to present
RESPONSIBLE AGENCY: 4950th Test Wing, Wright Patterson AFB and Edwards Air Force Base

THE STORIES

Through the years, the major example of using production aircraft to perform test missions has been accentuated by the operations of the 4950th Test Wing based at Wright Patterson Air Force Base and later at Edwards Air Force Base.

The organization maintained and modified a fleet of KC-135s, C-18s, C-141s, and C-130s of its own, and modified planes for other organizations, to perform a montage of test missions. Just about everything imaginable has been tested. Of interest is the age of many of these planes with a number of them dating back to the mid-1950s. Here is a listing of some of the more significant of these testbed veterans.

KC-135A (No 55-3122)

This aircraft has supported dozens of programs over the years since it arrived at the organization in 1958. The plane was initially modified for electronic countermeasures research along with the testing of new engine systems. In the 1990s, the aircraft was involved with the Airborne Imaging Transmission (ABIT) Program, which was managed by Wright Laboratory's Avionics Directorate.

KC-135A (No 55-3120)

The third KC-135 built, 55-3120 has always been an R&D aircraft. Its main mission over the years has been infrared signature analysis.

A major program was its conversion to the Flying Infrared Signatures Technology Aircraft (FISTA) to support the Infrared Properties Program. The aircraft was nicknamed "Steam Jet One" because of its water-injected powerplants. Programs supported included the Defense Advanced Research Projects Agency's (DARPA) Infrared Target and Background Program, the Air Force System's Command Teal Ruby Program, NASA's Hi-Camp Program, and the Strategic Air Command's Have Shaver Program.

The DARPA Program analyzed target background as they ap-

This KC-135A testbed aircraft tested new engine systems and was the ABIT test aircraft during its career. (Bill Holder Photo)

peared in space to serve as an aid in designing space detection systems. Teal Ruby was a test of the feasibility of using space-based infrared heat detecting systems to detect targets. Hi-Camp missions used a U-2 aircraft with an infrared detector flying a matching ground track at the same speed but at a higher altitude than the FISTA testbed. In Have Shaver, the FISTA KC-135 analyzed the effectiveness of camouflage on ground vehicles.

In 1988, the FISTA aircraft was involved in cruise missile support, F-16 infrared missions, and testing of a new infrared sensor. It was also active in infrared testing for the B-1B and F-15E aircraft. In 1990, the FISTA captured the infrared signatures of the KC-10 and KC-135R tanker aircraft. And most recently, the testbed collected infrared IR data on the F-117A and F-22 fighters.

KC-135A (No 55-3132)

The 15th KC-135 built, this aircraft has been modified over the years with large, top and bottom radomes and has carried out tri-service jamming and electronic countermeasures testing.

ARIA Test Aircraft

The most recognizable test aircraft in the world are the Advanced Range Instrumentation Aircraft (ARIA) airborne telemetry platform

The ARIA modification is the most recognizable testbed aircraft in the world. A number of Boeing 707s and KC-135s have been modified into EC-18s and EC-135s. This particular version is an EC-18 aircraft. (USAF Photo)

PART II: TESTBEDS

This photo shows the EC-18B ARIA configuration. The ARIA concept was developed in 1968. (USAF Photo)

with its bulbous nose design. This strange fleet includes both EC-135s and EC-18s.

The ARIA was developed in 1968 to receive, record, and relay telemetry data and voice transmissions between NASA astronauts and Houston Control during the Apollo moon program. Since then, many of these aircraft have been mission modified to support various DOD and NASA missions. ARIA involvement has included manned and unmanned space launches, cruise missile tests, Army and Navy ballistic missile launches, and Advanced Medium Range Air-to-Air Missile tests. Other ARIAs were modified to support the Sonobuoy Missile Impact Location System.

Of interest is the source of a number of the later ARIA aircraft when six commercial high-time 707-320 airliners were purchased from American Airlines in 1982 and designated C-18s. It marked the first time that such a commercial aircraft purchase was made.

KC-135A (No 12662) Five Satellite Communications Program
This particular KC-135 was modified in the late 1970s to accomplish simultaneous transmission tests with five satellites. The aircraft was equipped with multiple receivers that operated at various frequencies for the multi-satellite tests.

Besides the extensive on-board electronic additions to the interior of the aircraft, this plane was also characterized by the large black radar blister located on top of the fuselage just over the wing root. Over the years, this plane was also involved in a number of other test programs.

The LASERCOM modification of this C-135 testbed aircraft was used to transmit data between satellites and ground and airborne users. (USAF Photo)

107

ONE-OF-A-KIND RESEARCH AIRCRAFT

Above: The "5" in the tail number of this C-135 indicates 1955, which makes this plane, which served as the testbed for the Airborne Laser Program, one of the oldest of the type. (Bill Holder Photo)

Above right: This particular aircraft was easily identifiable from its pair of domes on top of the fuselage. (Bill Holder Photo)

Right: The Airborne Laser Program testbed aircraft now resides quietly in the Air Force Museum. (Bill Holder Photo)

C-135 (No 00377)

The LASERCOM (for Laser Communications) program was one of the biggest programs accomplished by the 4950th in terms of air-craft modification. Over a two year period in the late 1970s, over 200 Wing personnel worked to alter this aircraft's structure to flight test a laser designed to transmit phenomenal amounts of data between satellites and ground and airborne users.

The modification work included installing a laser window in the aircraft cargo door, an electro-optical (E-O) or laser assembly in the cargo compartment, an environmental shroud around the laser assembly, and five equipment racks with crew stations.

Most of the work on the LASERCOM C-135, though, involved the modifying of the cargo door to accept the special, 30-inch-diameter window dubbed "Eye of the Dragon" by its installers. The door essentially was torn down and rebuilt to accommodate the laser window, and once assembled, the door-window was pressure tested to insure safety of flight.

LASERCOM was a three-phase program which lasted into the early 1980s. The system demonstrated the flexibility of permitting lower-data-rate users to communicate between a greater number of spacecraft, airbourne, surface vessels and ground users.

C-135 (No 53123)

The Air Force High Energy Laser (HEL) Program, which was initiated in 1973, was supported by The Airborne Laser Laboratory (ALL), a highly instrumented KC-135.

Installation of the test laser required modification and redesign of a major portion of the aircraft, including the cutting of large holes in the floor and ceiling for the installation.

This program demonstrated that an airborne target could be destroyed by an aircraft-mounted laser. The modification was so extensive that when the program ended, the aircraft was retired. It is now on display at the U.S. Air Force Museum at Dayton, Ohio.

NC-141A (No 12779)

Becoming operational in 1990, the Advanced Radar Test Bed (ARTB) was designed to be an extremely flexible electronic counter-countermeasures test platform.

Major modifications to the nose of the early-production Starlifter allowed it to accept a number of interchangeable radomes including those of the B-1B, F-15 and F-16. The radar systems were installed by Lockheed Aeronautical Systems.

The testbed allowed tests and analysis of new radar system performance, and analysis of improvements to existing radars, not only in 'real time' but at low cost. Previously, radars were tested on the ground and then installed on the intended operational system.

According to officials associated with the ARTB when it became operational, this system represented the most advanced instrumentation systems available in the Air Force. It will be used to test integrated systems required for new aircraft into the next century.

The ARTB aircraft carries an onboard APG-63 radar in a compartment just behind the nose and a radar test instrument system in the rear of the aircraft.

KC-135 "Weightless Wonder" Testbeds

Long before men walked on the moon, astronauts floated across the cargo deck of this testbed KC-135, a plane known as the "Weightless Wonder." However, the people who flew them more often used the name "Vomit Comet" for obvious reasons. There were actually three of these aircraft so modified for the space simulation mission. The aircraft saw most of their action during the 1960s time period.

The simulation involved placing the planes into a shallow dive and then swinging into a 45 degree nose-up climb. As the plane

PART II: TESTBEDS

An earlier configuration of the Airborne Laser Program is shown during a flight test. (USAF Photo)

moved across the top of its arc, passengers experienced about 30 seconds of zero gravity. Besides human passengers, the planes also tested the effects of reduced-G on such equipment as lunar rovers and scooters.

Following the simulation duties, the planes were then converted back to other configurations.

C-130 (No 50022) ASETS Testbed

In 1986, major modifications were completed on this C-130 Hercules aircraft to carry an Airborne Seeker Evaluation Test System, or ASETS.

The ASETS system was used to test and evaluate current and future sensors and seekers used in air-to-ground missiles. It was similar in principle to a flying camera that takes pictures for missile guidance systems. It consisted of a 2,000-pound, precision, five-axis, retractable turret and several racks of electronic and data recording equipment.

Modifications to this Hercules included the cutting of a 58-inch diameter hole in the belly of the fuselage and the installation of clam-shell-type doors to close the hole. Inside, there was a special hoist system to lower a 2,000-pound turret into the opening after the aircraft was airborne, the turrets could be raised back into the aircraft prior to landing.

The turret extended 40 inches below the aircraft belly. The modification was considered major because three main-frame structural members of the aircraft had to be severed. Also, 400 pounds were added to the structural members to carry the added load.

The ASETS provided a capability for conducting competitive fly-offs of seekers/sensors by operating them side-by-side simultaneously under identical flight conditions.

This early model C-141 transport was modified for the Advanced Radar Test Bed (ARTB) and could be fitted with several different noses from current aircraft. (USAF Photo)

ONE-OF-A-KIND RESEARCH AIRCRAFT

One of the most famous KC-135 testbed aircraft was called the "Weightless Wonder" and served to train astronauts on the effects of weightlessness. (USAF Photo)

This C-135E testbed was modified to perform a satellite communications (SATCOM) mission and carried an abundance of optical equipment. (USAF Photo)

C-135 (No 00371) Have Lace Testbed

This particular C-135, under control of the Aeronautical Systems Division, was modified in 1986 by the 4950th Test Wing to perform the Have Lace laser communications experiments. The patterns visible on the side of the aircraft were used in previous experiments with the aircraft.

C-135 (No 00372)

In the mid-1980s, this C-135E was modified to accomplish a satellite communications test mission. In the program, which was managed by the Air Force Avionics Laboratory, the aircraft was modified with the installation of a large optical window in the cargo door, an optical radome on the top of the fuselage, a vibration-isolated mounting system for optical hardware and a microprocessor computing capability for automated data collection and in-flight analysis.

C-135 Open Skies Aircraft (No 12674)

In the early 1990s, a WC-135B-which in turn had been a KC-135B-was modified into the OC-135. The plane's mission was to support a treaty that established unarmed aerial observation flights over the United States, Canada, NATO and the former Warsaw Pact countries.

The aircraft modifications, besides its striking outside appearance, included one panoramic and three framing cameras. The modifications were completed in April 1993, and was followed by a series of flight tests. Other modifications included an auxiliary power unit, crew luggage compartment, sensor operator station, flight-following station, upgraded avionics, and compartments to maintain items for film.

The OC-135B can seat 38 people, including the cockpit crew, aircraft maintenance crew, foreign country representatives and crew members from the On-Site Inspection Agency (OSIA) located at Dulles International Airport in Washington.

C-141 Fly-By-Wire Testbed

Because of the interest shown in fly-by-wire technology and its potential application for the next generation aircraft, including transport aircraft, a unique C-141 testbed was developed by the 4950th test Wing to investigate the concept.

Although never modified to the point where the mechanical system was removed, this program demonstrated a capability where

The interestingly-painted fuselage of this C-135 participated in the HAVE LACE laser communications program. (USAF Photo)

PART II: TESTBEDS

The OPEN SKIES C-135 testbed aircraft is the newest version of the model for the 1990s time period. The mission of this sophisticated aircraft was to provide aerial observation to support an international treaty. (USAF Photo)

side sticks could be used in transports and also that command augmentation criteria and concepts and all fly-by-wire systems were practical and could provide the capability of achieving the same reliability as mechanical systems. Many of the concepts developed by this program were transitioned to the C-17 transport.

KC-135A (No 55-3128)

From a distance, this KC-135 might look like the standard Air Force transport version, but that would be wrong assumption. You see, this particular tanker (called an NKC-135) is a national asset for aerial icing and rain testing of both military and commercial aircraft.

But this testbed was not the first to perform the icing test function. In the late 1940s, a C-54 transport was modified for the icing mission. In the 1950s, aircraft so modified included a KB-50, KB-29 and KC-97, all managed by the Flight Test Division of the 4950th Test Wing.

Before the KC-135 was selected, there were also considerations in converting a B-47 jet bomber into an "ice plane." Eventually, though, this particular aircraft was selected, in addition to a C-130 transport.

In 1964, the KC-135 was modified to permit the simulation of aircraft icing conditions. The refueling boom was modified to spray water, rather than to pump fuel. Water droplets of every size found in nature can be produced. During tests, the subject aircraft flies closely behind the boom and gets sprayed with water from the NKE-135. This allows the test subjects' vulnerability to ice build-up and the effectiveness of their de-icing systems to be evaluated. The aircraft's P&W powerplants provided the source of bleed air used for water atomizing.

C-135A (No 60-0371) SDI Program Support

In 1986, the 4950th Test Wing operated an NC-135A called the Optical Diagnostic Aircraft (ODA) for the support of the Strategic Defense Initiative (SDI).

Due to the high security classification of the SDI program, many times Test Wing personnel were not informed of the exact purposes of the testing being accomplished. Most of the time, the crew flew the aircraft in flight patterns and operated the equipment in accordance with SDI directives without any other knowledge of the test itself.

The ODA aircraft also supported the SDI Delta 180 Test. For this mission, the ODA was one of three Test Wing aircraft which recorded data from the SDI spacecraft when it separated from the second stage of the Delta 180 boost rocket.

Finally, the aircraft was modified into the so-called ARGUS configuration where it was equipped with a high-resolution camera consisting of a 100-inch focal length telescope with a zoom video camera. Again, the nature of the testing didn't allow describing its mission.

The ODA testbed also supported the NASA space program, in particular supporting the space shuttle program. In 1986, though, with the loss of the Challenger, that portion of the ODA plane's mission was discontinued.

Earlier in its career, this particular aircraft supported the Airborne Laser Laboratory (ALL) program.

C-135 (61-2669) Speckled Trout

This KC-135, while assigned to the 4950th Test Wing, and called the Speckled Trout, was involved in many different test programs including the testing of autopilot systems, automatic navigation systems, radar evaluations, and in the mid-1980s, voice-activated control systems.

During the 1970s, the Federal Aviation Administration (FAA) and private industry used the testbed to acquire and reduce navigation information to study radical position errors. The testbed was also used to evaluate a number of navigation systems in the mid-1970s.

Other systems the Speckled Trout plane evaluated included the Ring Laser Gyro, an inertial navigation system that replace spinning gyros with a laser beam that continuously reflected a laser beam between internal mirrors, the Safe Flight Wing Shear program, a wing shear warning system, the Center of Gravity Fuel Level Advisory System, and the Auto Throttle System.

During 1988, Speckled Trout underwent the first of two aircraft modifications titled Transport Aircraft Avionics Cockpit Enhancement. Ultimately, the modifications resulted in a $42 million upgrade that included a Boeing 757/767 glass cockpit, a CRT-based engine indication and crew alerting system, a fully-integrated flight management system, and an auxiliary power unit. These upgrades formed the basis of the avionics architecture for the KC-135 avionics modernization program.

In 1992, the Speckled Trout aircraft was transferred to the Air Force Flight Test Center where it remains a viable testbed aircraft in its third decade.

It should also be noted that several other aircraft received the "Speckled" name tag with the Speckled Minnow (T-39, 62-4478) being used for research and test starting in 1974. In 1984, a C-21A (84-0098) became the new Speckled Minnow testbed, a mission it carried out until 1991.

C-131 Expandable Tire Testbed

During the early 1970s, a C-131 testbed aircraft was fitted with a modified main landing gear subsystem and expandable tires that were capable of being inflated and deflated in flight. The process was accomplished by compressors and pneumatic reservoirs mounted in the aircraft.

Expandable tire program. C-131 landing with expandable/deflatable tires. (USAF Photo)

The deflated tires occupied between one-half and two-thirds the space of a tire completely inflated, thus theoretically allowing considerable reduction in the wheel well storage space on an aircraft.

A number of problems were encountered during the testing causing the program to be terminated in 1971.

PART II: TESTBEDS

Ice Testing Testbeds

AIRCRAFT TYPE: Many types of multi-engine aircraft and one helicopter
MISSION: To provide icing testing on aircraft
OPERATING PERIOD: Late 1940s to present
RESPONSIBLE AGENCY: USAF 4950th Test Wing, Naval Air Warfare Center

THE STORY
Through the decades, a number of production aircraft have been modified with equipment to enable them to provide a spray of water at altitude on a trailing aircraft. Here are the major programs:

Early Attempts
In the late 1940s, a modified C-54 transport was modified with a propeller icing rig installed on it. There were, however, problems simulating the actual drop sizes produced by Mother Nature. During this same time period, some rather crude spray aircraft set-ups were also devised including a B-24 which produced its spray cloud by forcing water through a fire hose and fire hose nozzle. Also, a Lockheed Constellation used a two-inch pipe for a nozzle.

KB-29 Bomber
A more advanced spray system was installed on a KB-29 aircraft which consisted of a "T-Bar" arrangement at the end of a refueling boom. The bar had a series of 3/16 inch holes on each end of the T where water was forced into the airstream. The set-up was used mainly for simulation of crystal formation at high altitudes and for heavy rain at low altitude.

C-130 Transport
In the 1960s, the 4950th Test Wing used a C-130 transport to great advantage in icing experimentation. The aircraft was fitted with a sled device that provided a great improvement in icing simulation. There was also an arrangement to provide an adequate supply of hot air for air-water type nozzles to keep the spray rig free of ice accumulation. The main mission was the simulation of slow-speed icing conditions.

This KB-29 was equipped with in icing system mounted on its refueling boom. (USAF Photo)

KC-135 Tankers
During the same time period, there was also a modified KC-135 tanker that performed high-speed icing tests. The KC-135 was used for testing at speeds between 150 and 300 knots at altitudes below 30,000 feet.

The unique spray nozzle consisted of circular rings and cross bars of aluminum tubing which contained fuel injection nozzles. The design of the nozzle evolved through the years with efficiency increasing with each succeeding design.

A modification on another KC-135 used the aircraft powerplants to provide the source of bleed air used for water atomizing.

CH-47 Helicopter
Recent Navy tests have used the Helicopter Icing Spray System (HISS), a modified CH-47 helicopter. The HISS aircraft is considered more effective than the long-standing KC-135 icing aircraft because of the bigger water cloud it created.

This KC-135 icing aircraft lets loose a spray of icing fluid to check out the trailing aircraft. (USAF Photo)

The configuration of the icing mechanism on this KC-135 is clearly visible in this photo. (USAF Photo)

ONE-OF-A-KIND RESEARCH AIRCRAFT

KC-135 Winglet Testbed

AIRCRAFT TYPE: KC-135 Tanker
MISSION: To test potential performance enhancements from the addition of wingtip winglets
OPERATING PERIOD: Late 1970s to 1981
RESPONSIBLE AGENCY/S: Air Force Flight Dynamics Laboratory and NASA (Langley Research Center and Dryden Flight Test center)
CONTRACTOR: The Boeing Company

THE STORY

The goal of reducing large aircraft fuel consumption enjoyed a high priority in the 1970s. That goal was highlighted by a joint Air Force/NASA program to design, fabricate, install, and flight test a set of 'winglets' on an Air Force KC-135A testbed aircraft to reduce drag and thus demonstrate fuel conservation.

The KC-135 was chosen as the test aircraft because its wings could be easily modified and were of a design and sufficient size that data resulting from the program could ba applied to current and future transport-class aircraft.

The particular KC-135A selected for the testing was an ancient machine carrying the 1955 serial number of 55-3129. Those old birds just seemed to keep showing up to perform modern testing and research. Boeing fabricated and installed the winglets. The design of the winglets was based on research accomplished by Dr. Whitcomb of NASA Langley which was published in 1974.

The overall objectives of the program were to determine the effects of the winglets on the performance of the test aircraft and provide data for the selection of a winglet configuration to possibly retrofit the complete KC-135 tanker fleet.

Of particular interest was the fact that the winglet design used could be varied as to cant and incidence angles to optimize performance. The winglets could be changed on the ground between flights to any of three different positions.

The program was managed by the Air Force Flight Dynamics Laboratory with the initial goal of reducing the total aircraft drag at cruise by about eight percent. It was estimated at the time that drag reduction, when related to the entire KC-135 fleet, could be translated into an annual fuel savings of about 45 millions gallons.

Unlike wing end-plates, which had been suggested for a number of earlier years as a means of reducing drag, the winglets were designed with the same careful attention to airfoil design and local flow conditions as had been used on the design of the wing itself.

The sizable winglets were 9.33 feet in height, 1.92 feet at the tip, 6.07 feet at the root, and weight about 296 pounds each. The winglets were fabricated of aluminum and used standard aircraft construction techniques. The construction time for the winglets was about 16 months.

The winglets were instrumented with accelerometers, strain gauges and pressure taps. The aircraft and its engines were instrumented so that data such as aircraft loads, lift, drag and stability characteristics could be obtained in flight.

The Boeing-built winglets and outer wing panels were installed on the KC-135 testbed aircraft at the NASA Dryden Flight Research Center.

The maiden flight of the modified tanker took place on August 7, 1979 with a two-hour-plus flight where the testbed reached a speed of Mach .88 and an altitude of 21,500 feet. The innovative test program was completed in April of 1981 after a 215-hour, 55-flight test period. It was determined that the flying characteristics of the aircraft were basically unchanged and were in close agreement with preliminary analysis and wind tunnel tests.

The test results were analyzed by the Air Force before the decision was made not to retrofit the KC-135 fleet. It had been estimated that if the decision to go ahead had been made, that program could have started as early as 1983.

Flight test results confirmed predicted fuel savings although it was never stated exactly why the decision was made not to go ahead with the program, undoubtedly, economics was the main reason.

This artist's concept shows the installation of wingtip winglets on a KC-135 testbed aircraft. The fuel-saving concept was never incorporated on the Air Force fleet, but it's used extensively on the commercial airline fleet. (USAF Photo)

PART II: TESTBEDS

NASA/Langley Commercial Aircraft Testbeds

AIRCRAFT TYPES: Cessna 402B, Cessna Skyhawk, Beech T-34B, Poper PA-28RT, Gates Learjet
MISSIONS: A wide variation of missions
TIME PERIOD: 1970s through present
RESPONSIBLE AGENCY: NASA Langley Research Center

THE STORIES

Cessna Model 402B Businessliner
Arriving at Langley in 1982, this testbed has been extremely active performing a number of varying missions. In its first three years, it was involved with single-pilot IFR research. In the 1987-1988 time period, the 402B participated in ride quality studies. The final program, which began in 1989, involved the Shuttle Exhaust Particle Experiment (SEPEX).

Cessna Model 172K Skyhawk
The oldest of the Langley commercial fleet, this 172K Skyhawk came to Langley in 1972. The plane actively participated in a num-

This Cessna 402B testbed has performed a number of different test missions. It has been at NASA Langley since 1982. (NASA Photo)

The NASA Langley organization has a number of testbed aircraft including a Cessna 402B, 172 Skyhawk, T-34C, PA-28RT, and Learjet. (NASA Photo)

115

ONE-OF-A-KIND RESEARCH AIRCRAFT

This Skyhawk testbed has been serving since 1972. It's shown here carrying wing tufting, leading edge glove and twin instrumentation booms. (NASA Photo)

The T-34 performed laminar control studies in its tenure at NASA Langley. (NASA Photo)

ber of programs, but in the early 1990s was placed in non-flyable storage.

In 1975, the Skyhawk carried out pilot landing studies followed in 1979 with a series of handling characteristics tests. During the 1980s, the plane performed extensive stall/spin research.

Beech T-34C Turbo-Mentor
This testbed aircraft arrived at Langley in 1978 with its first major test program being laminar flow glove experiments. Later in the 1980s, the testbed was also be involved in the SEPEX experiments. The still-active-in-the-1990s aircraft also participated in the Vortex Detection Experiment in the 1989 time period.

Piper PA-28RT
Even though this aircraft arrived at Langley during the late 1970s, the testbed continues to be active in the mid-1990s. 1980s tests included stall/spin research, wingtip vortex turbine re-search, and in-flight wake vortex detection. During the 1990s, the programs involved infrared off-surface flow visualization, pressure belt measurements accuracy, and stagnation sensor evaluation.

Gates Learjet
This sharp-looking Gates Learjet arrived at Langley in 1984, and for its first five years performed laminar flow experiments. Then, during the 1990s time period, the Learjet was used to measure electric fields aloft, correlating with meteorological conditions to determine conditions conducive to triggered lightning.

The plane also participated in the Doppler Global Velocimeter (DGV) experiment.

This PA-2 testbed carries tufting on the wing, leading edge glove and twin instrumentation booms. (NASA Photo)

This Learjet performed laminar flow experiments. There are several portions of unpainted skin sections, not an unusual appearance on testbed aircraft. (NASA Photo)

PART II: TESTBEDS

L-100 High Technology Testbed

AIRCRAFT TYPE: Lockheed L-100-20 (Commercial C-130)
MISSION: Tactical Airlift Research and Development
OPERATING PERIOD: 1984-1993
RESPONSIBLE AGENCY: Lockheed Aeronautical Systems Company - Georgia

THE STORY

Modern tactical airlift is infinitely more complex than simply taking off and landing in short distances on rough forward operating bases. The mission may require an extended low-level flight over hostile territory with no navigation aids, followed by a cargo drop or landing on a short, bomb-damaged landing strip. The destination may not have any landing guidance and may be surrounded by hostile troops or threatening terrain. And last, the flight may also be at night and in bad weather. Obviously, this type of mission is not for just any crew, or just any aircraft.

Lockheed's High Technology Testbed, or HTTB, was built as a flying laboratory to help define and develop future aircraft for this demanding mission. It was used to determine requirements for future aircraft and crews, provide an airborne platform for developing the individual systems, and then integrate, test, and further refine the systems to make the mission performable and reduce pilot workload.

The HTTB was developed from a Lockheed Model 382E/L-100-20, a commercial stretched version of the C-130 Hercules. Lockheed developed the aircraft in conjunction with over 55 equipment suppliers who used the HTTB to further develop their own products. Developing a single piece of equipment to do a specific job is only part of the problem. Integrating each piece with all the other pieces of equipment and making them all work together as a system is an even greater challenge.

Lockheed made many airframe changes to allow the HTTB to better perform the tactical airlift mission. Major changes to the airframe included:

• More powerful engines and props

• High sink rate landing gear capable of absorbing a 10 foot per second sink rate following a no-flare landing

• Double-slotted flap for greater lift at low speed, capable of being retracted quickly after touchdown or for a go-around

• Extended chord ailerons for better roll control at low speed

• Wing spoilers to decrease lift for steep approaches and fast runway deceleration

• Cambered leading edge to allow a more nose-up attitude during steep approaches

• Horizontal stabilizer forward extensions to improve airflow over the tail when the flaps are extended

• Increased chord rudder to improve low-speed yaw control

The Lockheed HTTB testbed developed many new technologies for tactical airlift concepts. (Lockheed Photo)

ONE-OF-A-KIND RESEARCH AIRCRAFT

Many internal changes were also made to the HTTB to allow it to better perform the tactical airlift mission. These were primarily in the areas of the flight control system, avionics, and cockpit displays.

The standard flight control system was replaced with a digital stability and command augmentation system. This system made the aircraft more stable throughout its entire operating envelope, greatly reducing pilot workload and allowing approaches to be flown as low as 80 knots.

The digital flight control computers formed a triplex system that communicated with themselves and other electronic boxes through a fiber-optic communication link.

New high-performance actuators were installed on all the control surfaces. These units can be operated as standard hydraulic actuators, or in a "fly-by-wire" mode, receiving electrical drive signals directly from the digital flight control computers.

A "glass cockpit" using flat panel displays replaced most of the mechanical instruments used by the pilots. Advanced color graphics helped the pilots to visualize where they were, what was coming up, and other essential flight and navigation information. Flat panel displays were used because they are much lighter and smaller than older cathode ray tube displays. Much of this information could also be displayed on a Heads-Up Display, so the pilot could keep his eyes focused outside the cockpit.

Navigation was performed using a variety of systems, all integrated with the display system. These included three inertial navigation systems, a global positioning system, laser range finder, and forward-looking infrared radar. The infrared picture could be overlaid on the Heads-Up Display with graphical data to give the pilot a terrain picture at night.

HTTB also possessed an advanced data recording system to record every aspect of the aircraft's performance. Up to 1024 different signals could be recorded at 2600 times per second each. These data could also be telemetered to a ground station for simultaneous monitoring by engineers on the ground. For remote base operations, the HTTB carried a van fitted with telemetry equipment, recorders, and displays. Upon arrival at the operating location, the van was driven off, parked, and used as a mobile data monitoring and analysis center.

The HTTB performed many research programs to develop systems for tactical airlift between 1986 and 1993. On February 3, 1993, the aircraft was performing high-speed taxi tests at Dobbins Air Reserve Base in Georgia. New actuators and fly-by-wire controls were being tested. The number one engine was intentionally shut down as a planned part of the test to determine how well the automatic rudder control system would keep the aircraft going straight. The aircraft then went out of control while on the runway, briefly became airborne, then crashed and was destroyed. The tragedy claimed the lives of seven Lockheed flight test personnel.

PART II: TESTBEDS

PA-30 Twin Comanche Testbed

AIRCRAFT TYPE: Piper PA-30 Twin Comanche
MISSION: General Test Support
OPERATING PERIOD: 1967 - Present
RESPONSIBLE AGENCY: NASA Dryden Flight Research Center

THE STORY

Not all flight research needs to be performed on high-performance aircraft which are complex and expensive to operate. An excellent example of a "simple" testbed aircraft is NASA Dryden's PA-30, commonly called the Twin Comanche. This aircraft was purchased in 1967 and remained in service ever since, carrying the designation NASA 808.

The PA-30 originally was purchased for an experiment to investigate the handling qualities of general aviation aircraft. Most general aviation pilots give little thought to the effort that goes into making an aircraft "feel right." Most aircraft fly differently from one another, but with the exception of an occasional minor quirk, are still acceptable to most pilots. That occurs because by the time a new aircraft type is offered for sale, the design was refined to the point that the flight characteristics fit into acceptable ranges of values.

The first phase of this research effort involved an extensive analysis of the aircraft to determine all of its aerodynamic and control characteristics. To do this, the PA-30 first was sent to NASA's Langley Research Center in Virginia and placed in a full scale wind tunnel. It then flew a series of test flights to gather the same types of data in actual flight. The two sets of data were then correlated. This was the first time such a comparison had been done since the late 1930s. Many recommendations for desirable flight characteristics were made that were applicable to all similar size aircraft. Piper took advantage of the tests being done on one of their designs and used the data to make improvements in later production PA-30s.

The PA-30's career did not end here . . . in fact, it had just begun. With high performance military and commercial aircraft starting to use advanced electronic flight control technology to improve performance and widen the operating envelope, the PA-30 was used to study the application of this technology to general aviation aircraft. A yaw damper to decrease undesirable lateral motions was developed and tested, as was an angle-of-attack display to allow the pilot to control the aircraft's pitch attitude and airspeed more precisely.

A variable stability system was added in 1969 to allow a wide variety of flight characteristics to be duplicated. Two types of control mechanizations were evaluated for use in future aircraft: rate command and attitude command control systems. What this means is that as the pilot moves the control wheel, the computer commands either a rotation rate or an attitude proportional to the wheel deflection. The computer did this by measuring the control wheel deflection and moving the control surfaces just the right amount needed until the desired rotation rate or attitude was achieved. New military aircraft soon to enter production would include these new control mechanizations, and their potential for use in general aviation was studied.

In early 1972, the PA-30 was used to help develop short takeoff and landing (STOL) approach patterns. Future STOL aircraft would fly their final approach at much slower speeds that conventional aircraft. Since they still would have to operate often at regular airports, their slow approach speeds would make it difficult for them to mix with conventional aircraft at a busy airport. It was

The PA-30 Twin Commanche served since the 1960s as a general support testbed. (NASA Photo)

determined that a curved final approach for STOL aircraft, leveling out over the end of the runway, would best allow them to co-mingle with the long straight-in approach of conventional commercial aircraft.

For test purposes, the curved and descending flight path pattern was programmed on a computer on the ground. Using radar, the precise location and altitude of the aircraft was determined and the computer then calculated guidance signals to direct the aircraft to the desired ground track. These signals were transmitted to the aircraft and displayed to the pilot using glideslope and localizer steering bars. During the tests, the turn radius and descent steepness were varied to determine the best combination. Using the variable stability system, the aircraft's flight characteristics were varied to show that the approach path could be flown by aircraft with a wide range of flight characteristics.

In late 1972, the PA-30 began another test program, this time to help develop crew station requirements for remotely piloted vehicles (RPVs) in which the pilot flew the aircraft from the ground. This technology was being developed for military use and also for high-risk flight testing. Although RPVs were already in military use, the RPV pilot flew by controlling the autopilot, essentially specifying the heading, altitude, and airspeed to the system. New technology would allow the pilot, seated on the ground in a cockpit mockup, to fly with stick, throttle, and flight instruments, and fully control the aircraft just as if seated in the aircraft. High speed telemetry to transmit all pilot commands to the aircraft and then transmit aircraft flight information to the ground was developed, installed, and tested to demonstrate that such remote control worked and was usable. A variety of airborne cameras were mounted in the PA-30 so that a variety of focal lengths and fields of view could be tested. For safety purposes, a pilot rode in the PA-30, ready to take over if the system failed or if the remote pilot had difficulty.

In late 1974, a new landing program was performed to demonstrate that aircraft could be landed without any outside visibility. High speed flight places severe restrictions on window size and location. Thick window or canopies are very heavy and do not carry much structural load. Thus, the surrounding metal structure must be very heavy, further increasing gross weight. Future supersonic aircraft could be much lighter and stronger if there were no windows. Unfortunately, this would create a problem for the pilot!

For this test, a television camera was mounted on top of the fuselage and a 5 x 7 inch television screen was installed in the instrument panel facing the left-seat pilot. A cloth curtain blocked all outside visibility. With a safety pilot in the right seat, actual landings relying solely on instruments and the television picture were demonstrated. In further tests, certain flight control features were mechanized, such as automatic throttle control, to determine which control automation could best assist the pilot.

In 1984, the PA-30 was used to develop and test the instrumentation that would be used in the remotely-piloted Boeing 720 Controlled Impact Demonstration Program. The pilot who would fly the actual impact flight sat at the remote operator's station and practiced the approach profile using the PA-30.

NASA Dryden began a series of investigations in the use of thrust-only control in 1989. This followed an airliner accident in which all normal steering control was lost and the pilots flew the aircraft to a crash landing using only the engines. The idea is that all engines could be speeded up or slowed down to produce pitch changes, and operated asymmetrically to produce roll/yaw motions. Potentially, an emergency mode in the flight control computer could allow the pilot to continue flying using the control wheel, but the computer would command thrust changes to maintain some level of controllability. In a preliminary test using the PA-30, gross pitch control was demonstrated using thrust only, but landing would have been difficult. Further testing using the Calspan Learjet and a NASA F-15 demonstrated greater promise.

The veteran PA-30 remains an active and valuable test asset at NASA Dryden. It has shown over and over that valuable research and flight testing can be performed economically and reliably with small aircraft.

PART II: TESTBEDS

Sabreliner Supercritical Wing Testbed

AIRCRAFT TYPE: Sabreliner Model 60 business jet
MISSION: Development of a production supercritical wing
TIME PERIOD: Late 1970s
RESPONSIBLE AGENCY: Rockwell International company effort
CONTRACTOR: Raisbeck Group

THE STORY

The technology and benefits from a super-efficient wing design were getting considerable interest during the 1970s, both in the military and commercial arenas.

Rockwell International decided that the technology could improve the performance of its next business jet, the Model 65, so a program was initiated using a modified Model 60 to prove that the concept was viable.

The late 1970s effort involved the Raisbeck Group of Seattle, Washington accomplishing the significant modifications. The modifications included the fabrication of a production quality supercritical wing that included full-span blunt fixed leading edges replacing the existing leading edge flaps, along with chord increases at the wing root and tip, addition of wing stall fences, and Fowler flaps.

Also, the wing span was increased by six feet, three feet on each wing. The total aerodynamic package was called the Mark Five system.

But besides aerodynamics, there were also power modifications substituting Garrett Corporation TFE731-3 turbofan engines for the Pratt & Whitney JT12A-8 engines. Also added were Grumman Aerospace nacelles equipped with fixed-point, target-type thrust reversers.

The modifications added significant weight to the testbed aircraft, from about 20,400 pounds maximum weight for the Model 60, to about 24,000 pounds for the testbed. The testbed aircraft was built at Rockwell's Sabreliner Division in Los Angeles.

The Model 65 testbed made its first test flight in July of 1977. The test program lasted for about two years with much of the acquired technology incorporated into the follow-on Model 65.

ONE-OF-A-KIND RESEARCH AIRCRAFT

SR-71 Testbeds

AIRCRAFT: SR-71 Blackbird
MISSION: Used for test programs requiring high-speed aeronautical capabilities
TIME PERIOD: 1990s
RESPONSIBLE AGENCY: NASA Dryden Flight Research Center
CONTRACTOR: McDonnell-Douglas

THE STORY

The SR-71 Blackbird. It could well have been the most famous USAF aircraft ever built. With its outlandish design, cold sinister black paint scheme, unbelievable performance, and the fact it was always shrouded in secrecy, that reputation is easy to understand.

The SR-71 performed its Mach 3+ reconnaissance capabilities with great precision for many years and it looked like it would be around for many years. Then in 1990, the unbelievable happened, when it was announced that the Blackbirds were to be retired. The reason given was that the planes had become too expensive to support.

NASA quickly realized that having an aircraft with this capability would be a great testbed aircraft for a number of missions where high speed and high altitudes were required. For that reason, the space agency acquired three of the aircraft and has individually modified them for test missions. Although we're not exactly talking one-of-a-kind here, the testbed missions these planes accomplished are worth mentioning.

The planes have been used for a number of different missions including the testing of new materials and designs, carrying telescopes and other research instruments to the edge of space. NASA scientists' mouths are watering with the test possibilities of the aircraft.

One interesting test in the early 1990s involved the simulation of low and slow performance of a planned US commercial supersonic jetliner which might be decreasing from supersonic to subsonic speeds over a populated landmass. The SR-71 created multiple sonic booms across the California desert to examine the possible effects.

Another test involved the testing of a laser measurement system which was designed to provide pilots with more accurate data on altitude and airspeed. There were even more vigorous plans in the works where a simulated wind tunnel in the sky could be created by attaching models to the test aircraft.

Still other testing planned for the SR-71s involve their use as airborne science observatories carrying of cameras and instruments which could look both up and down.

Other applications of the unique trio include use as a research platform for simulating the characteristics of a spacecraft. Such a use would be considerably cheaper than using the real item.

Even though the Blackbird's roots reach far back into the 1960s, its capabilities still stun the brain. Each of its three ton powerplants pumps out 35,000 pounds of thrust which can propel the projectile-like aircraft to Mach 3.2 at over 85,000 feet.

Another planned test for the aircraft is the Optical Air Data System. This system incorporates the use of lasers to measure the velocity of particles in an air mass rather than the standard method of measuring static and pitot pressures. Along with the Optical Air Data System, the Jet Propulsion Lab planned to "piggy back" an experiment with an ultra-violet spectrometer. It will be used to measure ultraviolet atmospheric backscatter. The pro-gram was expected to last for six months.

For later in the 1990s, a number of requests had been made from several different commercial companies. Motorola is involved with a communications satellite experiment to test the communications link to the ground. Martin-Marietta has asked to have a microwave imaging system on board one of the SR-71s.

In 1994, the SR-71 came under consideration as the launching platform for targets for the next-generation missile interceptor system.

The tests call for one of the three SR-71 testbeds to carry and launch a target system mated to a Minuteman third state booster. In effect, the SR-71 would serve as the "first stage" of the airborne launching system platform.

And so it goes as these SR-71s could well be the most productive testbed aircraft ever.

Late flash! In 1995, the Air Force decided that the SR-71s were still needed and the three Blackbirds were returned. Who knows, maybe the birds will be converted back to testbeds sometime in the future.

The SR-71 was used for many years as a high-flying reconnaissance aircraft, but when it was recently retired, three of the Black Birds were transferred to NASA for research work as flying testbeds. (NASA Photo)

PART II: TESTBEDS

Boeing 737 Terminal-Configured Vehicle (TCV) Testbed

AIRCRAFT TYPE: Boeing 737
MISSION: To develop a number of new aircraft concepts
TIME PERIOD: Late 1970s to 1990s
OPERATING AGENCY: NASA Langley Research Center
CONTRACTOR: Boeing

THE STORY

This Boeing 737 testbed aircraft has been one of the most used research aircraft in NASA history investigating a number of cockpit concepts along with other developments. By the early 1990s, the craft had accumulated some three thousand hours of test time. The plane is interesting in that this particular 737 was the prototype test aircraft for the 737 transport.

The testbed was actually rescued from the 'junk yard' by Boeing and NASA, and it was converted into a flying laboratory. In 1974, it was moved to the Langley Research Center.

Every cubic inch is used in the testbed. At the front of the fuselage, there is a low light TV system. Further back, within the fuselage, is a second cockpit (called the research cockpit). In what is normally the passenger area, there is a maize of electronic gear, the key equipment being the Flight Control Interface, Image Converter, Data Acquisition System, and the Flight Test Engineers Station. No room for any paying passengers on this modified 737!

Over the years, NASA Number 515 has been used in a multitude of test programs. Its initial program (1975-1977) saw the TCV testbed support a Microwave Landing System Evaluation & Demonstration program.

Between 1975 and 1985, the TCV participated in a pair of display test programs-the development of a Velocity Vector Display and design of CRT electronic cockpit displays and formats, both concepts to aid pilots in accomplishing their job.

In 1983 and 1986, the programs addressed profile descent and total energy control law developments. In 1989, there was important work with helmet mounted displays for precision manual landings.

TCV research during the 1980s also showed that it was possible to place the airplane at a point in space, for example, the start of the descent stage. The TCV was able to accomplish that descent along a path that used minimum fuel. The technique also eliminated the need to stay in a holding pattern because the aircraft speed could be determined precisely so the plane could arrive at a specific location at a specific time.

NASA explained that the most recent work of the TCV was, "To develop procedures to improve the capacity and efficiency of air terminal areas, improve approach and landing capabilities in bad weather including wind shear, and reduce the exposure of communities to aircraft noise."

This standard airliner, with its large volume and performance capabilities, was a perfect choice for the missions it has performed through the years. The main portions of the trajectory for a majority of the testing include the descent, terminal area transition, approach, and turnoff.

A highly-modified Boeing 737 was designed as the Terminal-Configured Vehicle (TCV) which contained a complete second cockpit within the fuselage. (NASA Photo)

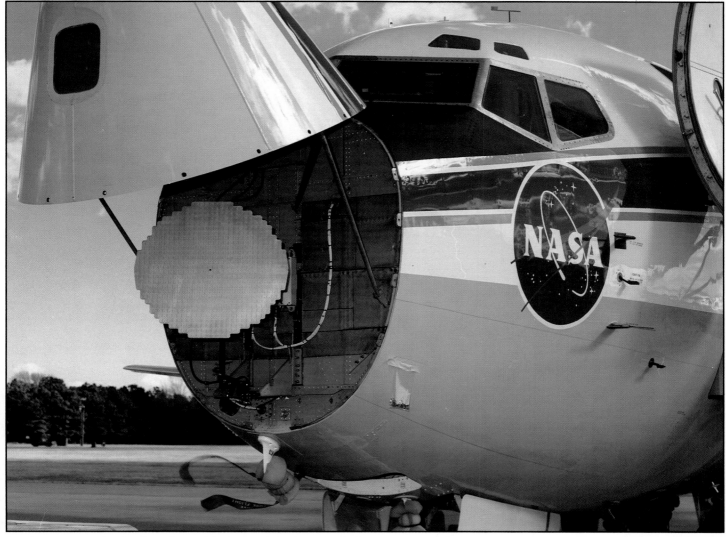

A radar is contained under the radome of the Terminal-Configured Vehicle (TCV). (NASA Photo)

When the 737 testbed is flying a test mission, the research pilots in the research cockpit are actually flying the plane, with the pilots in the conventional cockpit position serving as back-ups if anything should go wrong.

The need for the TCV capabilities was greatly accentuated in the late 1970s when delays were costing the airlines about a half-billion dollars annually. A suggested solution to the problem was the use of a time-controlled descent to the airport.

Adding a 'smart' airplane with advanced avionics systems would permit the controller to progress from active metering of traffic to passive metering. That's where the TCV program came and investigated the techniques needed to achieve the assigned-time objectives accurately and effectively.

The TVC was retired in 1994. It will soon be replaced by a Boeing 757 configured for similar research.

PART II: TESTBEDS

Boeing 720 Controlled Impact Demonstration Testbed

AIRCRAFT TYPE: Boeing 720
MISSION: Improved Airliner Safety
OPERATING PERIOD: 1984
RESPONSIBLE AGENCY: NASA Dryden Flight Research Center

THE STORY

Aircraft are designed to fly. Everything about them was placed just so to enable them to leap into and stay in the air. An aircraft's greatest beauty is seen in the sky when all parts are working as designed. When an aircraft has lived out its life and is no longer useful, it usually gets melted down to recover its metal or disassembled to salvage useful parts.

In the days of sail-powered warships, old veterans were usually towed out to sea and sunk with their colors flying, a final military tribute to an elderly warrior. A modern equivalent occurred in 1984 when an aged Boeing 720 airliner was intentionally crashed into the desert at Edwards Air Force Base. However, the purpose was not to end an airliner's career in a brief flash of glory, but to conduct scientific research to help make future airline crashes more survivable.

Airliners are among the safest vehicles in which to travel. They are many times safer than the family car. Airliners seldom suffer catastrophic failures in flight, even after encountering severe conditions such as storms. Most accidents occur at takeoff or landing. Many fatalities resulting from these accidents could be avoided if the aircraft's structure could hold together better and if post-crash fires could be reduced or avoided.

In a joint effort performed by the Federal Aviation Administration (FAA) and NASA's Dryden and Langley research centers, a crash test was performed to help make future airliners better able to survive crashes. The test would evaluate the use of anti-misting kerosene, or AMK, for use in airliners. AMK is a fuel that does not evaporate readily, and thus will not burn in the open air. In order to make it burn in the engine, special devices spray the fuel in microscopic droplets as it enters the combustion chamber. Thus, fuel spilled after a crash landing should not burn, virtually eliminating post-crash fires. The test also was to study various structural aspects of crash dynamics to determine how various components fail, and to look at different seat restraint designs.

To perform the test, the FAA provided a 23-year-old Boeing 720. It had been used as a training aircraft and had logged over 20,000 hours and 50,000 takeoffs and landings. For this program, it carried the designation NASA 833. Work began in July 1983. NASA Dryden engineers developed a remote control capability so pilots could fly the aircraft from a ground station for the actual test. NASA Langley developed an extensive instrumentation and data recording system. About 75 anthropomorphic dummies were seated in the passenger cabin. Each dummy had the same mass distribution as a person, so its motions in the crash would duplicate closely those of a person. Thirteen of the dummies were instrumented and connected to the aircraft's recording system. Several video cameras located in the passenger compartment recorded the dummies' motions during the crash.

The crash site was a specially-prepared gravel runway on a remote area of the dry lake bed at Edwards Air Force Base. Several wing openers, large structures designed to rip the wings open, were located at the beginning of the runway. In the actual test, the 720 was to land gear-up just short of the runway, hit the wing openers, then slide to a stop.

Several preliminary test flights were performed during late 1984 to test the remote control capability, practice approaches to the crash site, and insure that the test instrumentation and recording systems were functioning properly.

Above and two on the following page: The Boeing 720 CID testbed aircraft erupts in flames as it slides through the impact site at Edwards Air Force Base, California, the ultimate price a testbed occasionally has to pay. The sequence shows the airplane hitting left wing low, impacting the barriers and tearing off the right wing. It then erupted into a fireball, but the flames from the special fuel quickly died out on their own. (NASA Photo)

ONE-OF-A-KIND RESEARCH AIRCRAFT

PART II: TESTBEDS

NASA/Boeing 720 taking off on a test flight prior to the actual crash test. (NASA Photo)

The actual crash finally occurred on December 1, 1984. Over 400 news personnel observed the test from a safe location about four miles from the impact point. As the 720 descended toward the target, the big craft rocked slightly and the right wing impacted the ground first and snapped off. The rest of the aircraft then hit the wing openers and burst into flames. The nose rotated about 90 degrees as the aircraft slid on the gravel, finally coming to a stop. Despite the spectacular fire ball, most of the flames extinguished themselves within ten seconds. NASA 833's final seconds were repeated on the television news thousands of times that evening, making this test one of the most widely viewed aircraft flight tests.

ONE-OF-A-KIND RESEARCH AIRCRAFT

X-21A Laminar Flow Control Program Testbed

AIRCRAFT TYPE: WB-66D
MISSION: Testbed aircraft for study of laminar flow control for large aircraft
OPERATING PERIOD: 1960 to 1964
RESPONSIBLE AGENCY: U.S. Air Force Air Research and Development Command
CONTRACTOR: Northrop Corporation

THE STORY

Through the years, there have been a long series of X planes, but the two X-21A aircraft were the only X craft that were not built from scratch to be a test aircraft. In this program, the testbed aircraft were a pair of highly-modified WB-66A USAF bombers, numbers 55-0408 and 55-0410.

The purpose of the program was ambitious to say the least with a number of significant goals. Possibly the most important was to acquire data for following generation aircraft which could use laminar flow control (LFC) concepts.

The idea of laminar flow aircraft had been investigated since the 1930s with the goal to smooth the flow over the surface of the wing. A number of different techniques had been investigated with the Germans investigating the concept during World War II. An F-94 would later be modified to investigate the LFC concept in this country.

The benefits of possessing an effective laminar control aircraft are measurable. For example, Northrop estimates on this X-21A showed that the break-even payload for the aircraft would be 8000 pounds lower than the normal configuration of the plane.

An operational version of an LFC aircraft would have take-off characteristics similar to normal planes even though there exists a 25 percent lower thrust because of the lower wing loading. Granted, the fabrication of an LFC system for an operational aircraft will end up costing more, but its flight efficiency would quickly compensate for that deficiency.

The ambitious X-21 program began in 1960 when Northrop was contracted to modify two WB-66 bombers into the X-21A configuration. The changes were significant as the aircraft were fitted with 30 degree swept-back wings incorporating so-called laminar control suction slots. The slots were practically invisible being only several thousandths of an inch wide.

Compressors, located in the nacelles, provided the sucking power to remove the boundry layer air on the wing surface. In order to aid in the investigations, the engine location on the planes was changed to the rear fuselage position.

Finally, atop the fuselage at the wing-joining position, there was a 'camel's hump' which aided in the flow pattern. It was hoped that the turbulence created by the air flowing over the wing could be reduced by 50 percent.

The first X-21A took to the air in May 1963. The second aircraft, which would fly in August of the same year, would be basically the same aircraft with a few minor subsystem differences. The second plane (55-0410) also remained in bare metal throughout the entire program.

The flight testing proved that boundry layer control was a vi-

The X-21A was the only "X" aircraft that wasn't built from scratch. Two of them were built from a pair of B-66 bombers. (USAF Photo)

PART II: TESTBEDS

Not all testbeds end up gloriously. One of the X-21A testbeds ended up deserted on the desert. (NASA Photo)

able concept with a measurable reduction of the friction drag. The effect was found to be more effective on the inboard portion of the wing. There were also leading edge wing fences added to improve the airflow. Final test results showed that the LFC effect covered approximately three-fourths of the top and bottom sides of the wings. It was certainly evident to test pilots when the LFC system was functioning, and a considerable degration in performance occurred when it wasn't working.

But just as the LFC was starting to show real promise, it was basically overtaken by technology with more fuel efficient jet engines. There was, however, serious consideration given to applying the effective principle to the C-141 Starlifter transport. Preliminary estimates indicated that an LFC system installed on a C-141 could increase its payload by 74 percent, or its nonstop range by 50 percent. Unfortunately, this prediction never had the opportunity to be proved.

ONE-OF-A-KIND RESEARCH AIRCRAFT

YF-23 Loads Study Testbed

AIRCRAFT TYPE: YF-23 Advanced Tactical Fighter Prototype
MISSION: To study loads calibration techniques
OPERATING PERIOD: 1994 to ?
RESPONSIBLE AGENCY: NASA Dryden Flight Research Center
CONTRACTORS: Northrop and McDonnell Douglas

THE STORY

As was the case with the earlier prototype YF-16 (which was converted for a number of test purposes), the YF-23 would also find a test function in the early 1990s. The purpose of the testing would be to study loads calibration techniques.

Actually, there were a pair of YF-23s built, but only one of the advanced fighters will be utilized in this testing. The aircraft, which was developed jointly be McDonnell Douglas and Northrop, was provided to NASA Dryden under a gift agreement which was signatured by both NASA and Northrop.

Before the testing could begin, though, it was necessary to completely 'declassify' the plane. During the test program when it competed with the winning YF-22 in the Advanced Tactical Fighter (ATF) program, the plane did carry a number of classified systems and subsystems.

Following the ATF competition, the two prototypes would placed in storage with no particular future mission planned. But as is often the case, this test mission did come up and the YF-23 was exactly the aircraft needed to fulfill the test requirements.

Before being placed in storage, the planes were completely stripped including powerplants along with all electronics.

The project was interesting with the goal being to improve the accuracy of current strain gauge load calibration methods. Strain gauges are used to measure loads on the aircraft structure and components. Data produced from ground tests are used to derive shear, bending moment, and torque loads equations. Once flying, strain gauges installed throughout the structure are used to help determine aerodynamic loads. Flight loads measurement is an important element of flight safety and aircraft certification and this project will really aid in accomplishing that goal.

The YF-23 was perfect, as if built from scratch to fulfill this test mission, because of its composite construction representing the latest in aircraft construction and aircraft materials. For comparison, a NASA F-16 testbed, which used more conventional fabrication techniques and materials, was used.

As of press time, the program was planned for a 1994 start with the test effort expected to last "three or four years" with the final results openly published and made available to other gov-government agencies and the aerospace industry in general.

Undoubtedly, in the years to come, these two prototype aircraft will serve in a number of test missions.

This photo shows the pair of YF-23 prototypes just after they were removed from storage. One of the craft would eventually be used to study strain guage loads calibration techniques. The other, for the time being, will remain in storage. (NASA Photo)

PART II: TESTBEDS

A strange new aircraft to serve as a testbed aircraft is one of the YF-23 prototype fighters which will be used in loads testing. It's shown here in its roll-out. (USAF Photo)

ONE-OF-A-KIND RESEARCH AIRCRAFT

Miscellaneous Testbeds

This F-5E testbed aircraft tested a composite material landing gear strut in 1987. This particular F-5E was a member of the USAF Aggressor Squadron, another example of a test being accomplished as a part of the normal operational mission. (USAF Photo)

This A-7D testbed, along with seven other operational versions, was fitted with composite material outer wing panels. (USAF Photo)

PART II: TESTBEDS

Increased maneuverability was the goal of this A-7 testbed modification with the addition of leading and trailing edge flaps. (USAF Photo)

A highly-instrumented F-4 testbed was tested to determine how much surface roughness the plane could operate from. The program was called "Have Bounce." (USAF Photo)

ONE-OF-A-KIND RESEARCH AIRCRAFT

This F-14A testbed was modified with the so-called Aileron-Rudder Interconnect System which investigated new high angle-of-attack flight techniques. (U.S. Navy Photo)

F-16 testbed aircraft fitted with a LANTIRN navigation pod during early testing of the system. (USAF Photo)

PART II: TESTBEDS

The longstanding NASA F-106 testbed on a test flight. Note the black appendage on the top of the rear fuselage performing some unknown test function. Also note the lack of military markings and the civilian registration number on the aft fuselage of this NASA testbed aircraft.

Would you believe this F-80 testbed? This unique testbed investigated the concept of a prone-lying pilot, hence the forward-located second canopy. (USAF Photo)

This busy USAF F-106 testbed was involved in many programs, including an early Integrated Fire Flight Control (IFFC) program, and the fitting of a M61-A1 Gatling Gun on the bottom of the fuselage. (USAF Photo)

PART III
PROTOTYPES

Besides the more prevalent simulation and testbed aircraft, there is one other category of one-of-a-kind research aircraft which must be included in this book to tell the complete "One-Of-A-Kind" story.

What we're addressing in this chapter is the modifying (sometimes major modifying) of a production aircraft as the prototype of a completely-new aircraft. It's happened a number of times.

Needless to say, use of already-existing parts and pieces greatly reduces the cost of fabricating such aircraft. In this era of immensely expensive aircraft systems, this technique could be employed more and more in future years.

There have been examples of both military-sponsored, and company-sponsored, efforts in this arena.

There is, however, one slight qualification which must be made for this category of "one-of-a-kind" aircraft. There isn't always one prototype built, sometimes there's a pair or trio, but the principle is the same. Hence, these concept aircraft are included.

PART III: PROTOTYPES

YA-7F (A-7 Plus) Prototype

AIRCRAFT TYPE: A-7D Corsair II
MISSION: Prototype for new fighter
OPERATING PERIOD: 1988-1990
RESPONSIBLE AGENCY: Aeronautical Systems Division, Wright-Patterson AFB, Ohio
CONTRACTOR: LTV Aircraft Products Group

THE STORY

The idea was simple. Take the already-proven A-7 Corsair II close support aircraft, greatly increase its ordnance delivery and thrust capabilities, and produce a low-cost aircraft to fulfill the Close Air Support (CAS) mission. It was called appropriately, the A-7 Plus!

The plans called for addition of the LTV-developed Low Altitude Nigh Attack (LANA) system for night attack capabilities in addition to other modern additions including FLIR provisions, automatic terrain following capabilities, a head-up display (HUD), and a new weapons delivery computer.

Modifications were equally significant to the airframe featuring both front and rear fuselage stretches, wing strakes, an extended vertical tail, trailing edge flap augmenters, lift/dump spoilers, engine bay tail cone, and automatic maneuvering flaps. These modifications were specifically designed for air-to-ground missions and forward-based, short field operations.

A new, more-powerful engine greatly increased the aircraft's load-carrying capability along with its agility. The increased volume of the A-7 Plus also provided for more fuel needed by the new powerplant. There was also an airframe-mounted accessory drive unit for self-contained ground operations. A new electrical and environmental control system was designed to accommodate the avionics upgrades.

The design goals of the upgraded fighter were ambitious to say the least. The plane was to (1) be able to arrive combat ready without tankers for 24-hour attack capability, (2) possess improved ground acquisition, with up to a 17,380 pound multiple ordnance load, (3) demonstrate high performance and threat avoidance, and (4) accomplish an impressive targets-killed to aircraft-lost ratio. Tough requirements to be sure, but there was high confidence by the contractor that it could be accomplished.

The $133 million contract was awarded to LTV in May 1987 to construct two of the prototypes. Pratt and Whitney supplied the F100-220 engines which would power the testbed fighter.

The first YA-7F prototype made its initial flight in 1989 from LTV's Dallas facility. Piloted by LTV test pilot Jim Read, the aircraft flew for one hour and ten minutes. Other flight testing was

Here's the way the modified Corsair II fighter looked after its considerable changes and alterations. The modification was considered as a candidate for the Close Air Support (CAS) mission, but it wouldn't be selected. (LTV Photo)

ONE-OF-A-KIND RESEARCH AIRCRAFT

A Corsair II, like this operational U.S. Navy version, was selected by the Air Force for significant modification as the YA-7F Prototype Fighter. Air Force versions were used for the actual modifications. (U.S. Navy Photo)

The YA-7A took on a new design grace. Here's a shot of the prototype during flight testing. (LTV Photo)

PART III: PROTOTYPES

Production of one of the two YF-7F prototypes at LTV Aircraft Products Group. (LTV Photo)

carried out into 1990 at the Air Force Flight Test Center, Edwards Air Force Base, California.

The ultimate goal of the program was to upgrade as many as 335 A-7Ds which were already in the inventory. The price for the modification was basically half that of a new aircraft, or about $10 million.

But the A-7 Plus, which had so much going for it, would never come to fruition. The main reason was probably the age of the A-7 airframes which DOD officials said would cause problems.

ONE-OF-A-KIND RESEARCH AIRCRAFT

F-16XL (F-16E) Prototype

AIRCRAFT TYPE: F-16 Fighting Falcon
MISSION: Prototype for air superiority fighter
OPERATING PERIOD: Early 1980s through mid-1990s
RESPONSIBLE AGENCY: Aeronautical Systems Division (ASD)
CONTRACTOR: General Dynamics/Fort Worth

THE STORY

Although the F-16XL still carries the F-16 identification in its designation, this aircraft was a far cry from its original Fighting Falcon brothers. Devised in the early 1980s, the testbed aircraft was designed from the offset to be an advanced version of the F-16 incorporating new aerodynamic and systems technologies.

The vastly-modified aircraft, which started from a basic F-16 airframe, featured a highly-swept, cranked-arrow delta wing developed by the Fort Worth Division during years of intensive collaborative work with NASA.

There were actually two XL models built, using two fleet F-16s provided by the Air Force. One was a single seater, while the second employed dual seating. General Dynamics put its best team on the job, and the prototypes produced met all design requirements.

In addition to the considerable structure changes, the pair of XLs used both the P&W F100 in the single place version, while an GE F110 was the powerplant of choice in the two-seater.

Details of the modification reveal that the new wing had an area of 646 square feet, more than double the standard F-16. Graphite polymide composite wing skins replaced aluminum, providing the strength and stiffness necessary for maximum wing performance. In addition, the standard fuselage received a stretch of 56 inches, which when combined with the new wing, dramatically increased the fuel capacity of the radical new craft by 82 percent over the original. There was also an additional 40 cubic feet of volume for avionics and sensor growth.

The XL dramatically extended the F-16's capabilities as it was able to take off and land in only two-thirds the F-16's normal distance. It also demonstrated twice the bomb load and 45 percent greater combat radius for both air-to-air and air-to-surface missions.

Studies indicated that the XL could be armed for the air combat role with eight AMRAAM missiles. In the ground support role, the ordnance included Mk 82 bombs and tactical missile dispensers.

Both models were tested, and then were considered for possible production as the so-called F-16E version. But first, the model had to compete for that distinction with the new F-15E. Unfortunately for the General Dynamics entry, the F-15 version came out on the long end of the competition, and ultimately was mass produced. General Dynamics indicated that the plane could have been in mass production by 1986 had the contract been won.

Thus, the F-16XL story appeared to be over. Hardly! The radical-appearing prototypes would continue to serve in research and development roles for many years, probably stretching to the turn of the century.

In 1989, both XLs were brought out of storage for use in a flight test program to evaluate concepts designed to improve supersonic flight. They were transferred to NASA and ferried to their new home at Dryden Flight Research Center at Edwards AFB, Cali-

It might not look much like the F-16 from which it was derived, but the bat-appearing F-16XL started from a standard F-16 production aircraft. (USAF Photo)

PART III: PROTOTYPES

The F-16XL is shown on the far right, sitting among a number of different F-16 variant aircraft. (USAF Photo)

fornia. The aircraft were equipped with an experimental wing glove perforated with thousands of tiny laser-cut holes connected to an air pump located in the fuselage.

The purpose of the arrangement was to accomplish continuous air flow over the wing during supersonic flight. The obvious result was reduced drag. The laminar flow research could be an important step to increase flight efficiency and reduce fuel consumption of future high-speed transports.

The testbed aircraft were also involved with the SR-71 testbed in evaluation of sonic booms.

The number one single-seat F-16XL (No 849) acquired a striking black and yellow paint scheme in 1993 and performed a number of new tests. A number of wing add-ons were added to lend extra lift during take-offs for mid-1990s tests. The research involved two different types of high-lift flaps and suction devices that drew boundry-layer air through wing-skin perforations.

Undoubtedly, these two advanced birds will see more research work in the development of new fighter aircraft.

In 1993, NASA used one of the F-16XL testbed aircraft to support the agency's High Speed Research (HSR) program. For the test, the craft received a dramatic new black and gold paint scheme. (NASA Photo)

ONE-OF-A-KIND RESEARCH AIRCRAFT

F-16/79/101 Prototype

AIRCRAFT TYPE: F-16
MISSION: Prototypes for re-engined F-16 models, the F-16/79 and F-16/101
OPERATING PERIOD: Late 1970s-early 1980s
RESPONSIBLE AGENCY: General Dynamics (F-16/79) and USAF (F-16/101)
CONTRACTOR: General Dynamics/Fort Worth

THE STORIES

F-16/79

The need was obvious. The Third World fighter market was calling for a low-cost replacement for its fighter fleets. General Dynamics recognized the potential and felt that it had an answer to the situation.

Early availability and low cost were the two requirements for such a system, and the company felt that it had the answer with an F-16 variation, a company-sponsored program.

The new aircraft would use the basic F-16 airframe, but be powered with a lower-thrust J79-GE-17X powerplant (18,000 pounds thrust) as compared with the standard Pratt and Whitney F100 (25,000 pounds thrust) powerplant. The company, during the late 1970s design time period, saw a potential for as high as a five hundred sale possibility. An advantage of this concept was that the -79 versions could have been produced along with the standard F-16s on the same production line. The aircraft to be replaced included Northrop F-5s, F-4 Phantoms, F-104 Starfighters, and several other early-model foreign aircraft.

An advantage of this concept was that many of the countries targeted already had maintenance and logistics systems in place for the J-79 engine which also powered the F-4 and F-104.

Several changes were required for the new configuration. The most noticeable was the lengthening of the fuselage by some two and one-half feet to a 48.02 foot overall length. Also, the air intake had to be modified to accommodate the new powerplant. Included in that latter change was a fixed compression ramp to reduce airflow and increase pressure recovery. The modification proved to actually be more effective than expected, with only minor changes required to the J79 powerplant itself.

After all was said and done, a new aircraft was designed and configured from an existing production aircraft, in this case, an F-16B. The new craft, when completed, ended up weighing about 1,300 pounds more than the original F-16.

Flight testing in the early 1980s proved that performance estimates were right on the button for the -79. As expected, climb performance, sustained turn capabilities, and range of the new downgraded fighter were less than the standard Fighting Falcon, but still met the needs of the targeted customers.

The formal flight testing program of the F-16/79 was completed in mid-1981, logging 71 flight hours in 76 flights over a two month time period. The testing showed that longitudinal stability, acceleration, and cruise performance were actually better than predicted.

Everything looked good for this son of Fighting Falcon, but it was not to be. The export F-16 generated very little interest among potential foreign customers.

The reason for the rejection? Well, it was quite easy.

All those countries wanted the full-up F-16 carrying the standard F100 powerplant. Even though the performance of the -79 was impressive compared to existing third country aircraft, it still just wasn't good enough. So the program effectively died after the rejection.

Designed as a possible third-country fighter, this modified F-16 (known as the F-16/79) carried the J-79 powerplant. (USAF Photo)

PART III: PROTOTYPES

F-16/101

Where the F-16/79 Program was basically a downgrading program, the so-called F-16/101 Program was an engine substitution which could provide increased performance. The program involved the replacement of the existing powerplant with a General Electric F101 powerplant. The program was accomplished in 1981 under the Derivative Fighter Engine (DFE) program.

One of the very early F-16s was used in this research program. The testbed aircraft, SN 50745, had been the first F-16 released to the Air Force for Operational Testing and Evaluation (OT&E) many years earlier.

Development of the F101 powerplant began in 1979 and included extensive durability testing in preparation of the testing which started in December of 1980.

Program Manager Mike Calson at the time, from the USAF Aeronautical Systems Division, explained that the engine testing proved that the F101 engine was a viable propulsion system for the F-16. "In a relatively short time, we acquired a complete characterization of the F101 DFE in the F-16 including operability, compatibility, installation effects, and system performance. Everybody associated with the F101 program was impressed with the engine system. The pilots liked its throttle response, and the engineers were impressed by its durability and operability."

The overall F-16/101 program consisted of 58 flights for a total of 75 flight hours. During the flight test program, the engine was put through a series of tests such as throttle usage and performance testing, including sustained turns, military and maximum power climbs.

In a similar experiment to the F-16/79, an F-16 was also modified to carry the F101 powerplant. Again, the concept would not be adopted for the F-16/101. The brightly-colored aircraft prototype is the middle aircraft in this formation. (USAF Photo)

ONE-OF-A-KIND RESEARCH AIRCRAFT

P-51 Mustang-Based Enforcer Prototype

AIRCRAFT TYPE: World War II P-51 Mustang
MISSION: To serve as the prototype for a 1970s-style close support aircraft
TIME PERIOD: Mid 1970s to early 1980s
RESPONSIBLE AGENCY: U.S. Air Force
CONTRACTOR: Private company headed by David B. Lindsay and Piper Aircraft

THE STORY

To many, it was about as crazy an idea as you could imagine, but to developer David Lindsey, it was dead serious business.

Here was the idea. It's a well-known fact that the North American P-51 Mustang was considered to be the best fighter of World War II. The model also continued to be employed at the start of the Korean War. Then it would serve with a number of third world countries throughout the world. Finally, modified versions of the aircraft would fly in the Reno Air Races and remain in operation by a number of private owners. As far as everybody was concerned the Mustang's military career was over.

But Lindsey thought that the old bird still had potential for a modern battlefield in a greatly modified version. He called his new plane the Enforcer and used an old Mustang as the starting point for the new plane.

The biggest difference in the new Mustang was the powerplant. The old reciprocating engine was replaced by an Avco Lycoming T55-L-9 gas turbine engine. Lindsey felt, that in addition to the greatly-increased performance, the engine would also emit a small residual exhaust which would provide a low infrared signature for heat-seeking missiles. Inside the wings were six .50 calibre guns which were well hidden.

There were a number of modern modifications planned for the Enforcer including all-aluminum alloy construction, solid state avionics, improved pilot extraction capability, redundant controls, increased maneuvering capabilities, and a response time of 26 minutes for a distance of 150 miles.

The Enforcer was expected to have a gross weight of almost 14,000 pounds with the capability of hauling over two tons of ordnance.

Lindsey had become interested in Mustang modifications when he purchased his initial F-51D. He then formed his own company, Cavalier Aircraft, and began rebuilding the Mustang for a civilian market. He explained that he had also built a number of advanced Mustangs known as Cavalier Mustangs for the Military Assistance Program. Indonesia was one of the recipients of those modified Mustangs.

Then came the long battle with the services to test the new aircraft. The Air Force felt, though, that the plane would be very susceptible to a single-round kill. Lindsey argued that the key selling points of the Enforcer were its small profile and low-in-flight noise level permitting high-speed, low-level approaches with little warning to enemy forces. He pointed to the advantages of the Enforcer compared to the A-10 which the Air Force wanted badly. Finally, there was the advantage that the Enforcer was estimated to cost about one-fourth that of the A-10.

It wouldn't be until the early 1980s that the Air Force would finally agree to test the Enforcer. To that end, it awarded a contract to Piper Aircraft to conduct the $12 million test and evaluation program.

But as is well known, the Enforcer would never be accepted for production even though it did have a lot of desirable traits. It did, though, finally get a fair chance to be tested.

The Enforcer will long be remembered as the dream of a talented designer that thought the old P-51 still had a lot of fight left. He was almost right!

It might look a lot like a World War II P-51 fighter, but this testbed aircraft was developed from the old fighter for modern counterinsurgency missions. (Bill Holder Photo)

PART III: PROTOTYPES

The major difference in the "old and new" Mustang was the powerplant, which was changed to a turboprop for the Mustang Enforcer version. (Bill Holder Photo)

ONE-OF-A-KIND RESEARCH AIRCRAFT

Prototypes for Gunship Programs

AIRCRAFT TYPES: C-47, B-57, C-131, C-130
MISSION: To develop effective side-firing capabilities from circling aircraft
TIME PERIOD: 1960s and 1970s
RESPONSIBLE AGENCY: USAF
CONTRACTORS: General Electric

THE STORY

There have been a number of gunship programs over the years using a number of different aircraft, a majority of them using transport-type aircraft. Here are the major programs through the years.

C-47-based Program

During the 1960s, possibly the most famous of the side-firing programs was undertaken. Using a standard C-47 transport, the aircraft was equipped with gatling guns protruding from the windows and doors. This initial effort proved to be very effective in the Vietnam War, and was given the name of "Puff the Magic Dragon." It was just the first of several other programs to come.

The idea of making a gunship from a slow-flying transport was conceived by an Air Force officer, Major Ronald Terry, who had served a tour of duty in South America. He had noted that mail and supplies were sometimes lowered into remote villages in a bucket tied to the end of a long rope suspended from the cargo door of a slow-flying aircraft. As the plane circled in a steep pylon turn, the bucket tended to remain in one spot over the ground.

Terry envisioned a line of fire in place of the line of rope. Side-firing guns in the transport could be aimed by the pilot and kept bearing on a ground target as long as the plane orbited in a steep bank overhead. Then, when he visited Vietnam and noted the requirements there to carry on a limited war, the idea became more of a concept.

In an interesting experiment, the testbed C-47 Gunship actually underwent some of its testing in an actual wartime environment in Vietnam. In initial combat tests, the aircraft was used in conjunction with the First Air Commando Squadron in the defense of 19 outlaying forts manned by the South Vietnamese.

Three GAU-2/A Gatling-type machine guns were mounted aboard the test aircraft. Each of the guns had six rotating gun barrels and had the capability of firing 6,000 rounds per minute. The guns were manufactured by the General Electric Armament Division, Burlington, Vermont, while the gun mounts were fabricated by the Aeronautical Systems Division experimental shops at Wright-Patterson AFB.

AC-130-based Program

Starting in 1967, the 4950th Test Wing at Wright Patterson AFB was called upon to demonstrate side-firing capabilities from a C-130 type aircraft as a follow-on to the C-47 program.

The C-130 Gunship was developed using a C-130 testbed aircraft at the 4950th Test Wing at Wright Patterson AFB. (USAF Photo)

PART III: PROTOTYPES

This photo shows the mini-gun and cannon which are located above, and behind the landing gear, respectively. (Bill Holder photo)

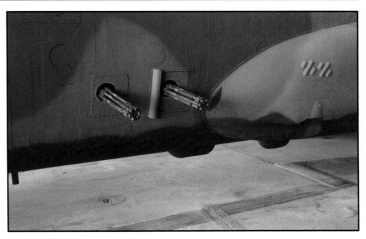

Shown here on the C-130 Gunship testbed are the set of Gatling Guns located just ahead of the landing gear fairing. This aircraft can be viewed at the Air Force Museum. (Bill Holder Photo)

A C-130A was modified for this new mission and tested at Eglin Air Force Base. The tests involved use of a 40mm gun, special ammunition, and an inertial targeting system. The purpose was to determine weapon accuracy, munitions effectiveness, and to improve the aircraft fire control system. There was some damage to the aircraft during the ground firing of the 40mm gun. It was determined that the damage occurred due to the effect of the static conditions of no air flow and the effect of the concrete ramp under the aircraft. Inflight tests of the system, though, proved that the concept was feasible.

Shortly later, a follow-on program was conducted on the so-called AC-130 Gunship II Follow-On program. The additionally-modified configuration incorporated a Forward Looking Infrared (FLIR) system to locate and track targets rapidly and a computer system to compensate accurately for wind and offset variations.

The Gunship II configuration carried four 20mm cannons and four 7.62 machine guns. Surprise Package was the name given to another AC-130A test program. This particular AC-130A incorporated the Pave Crow sensor to locate potential targets on the ground. The system was ultimately deployed to Southeast Asia where it proved to be an extremely effective weapon against ground targets.

In 1971, a program was initiated to install a large calibre gun on a prototype AC-130. The 4950th tested the feasibility and flight tested the installation of a 105mm howitzer on the transport. The testing also involved the use of ECM and flares to provide improved gunship protection. Again, the tests proved to be so effective that the technology was quickly deployed to southeast Asia.

In addition, there was also an AC-130E that was modified with new engines and several experimental items. Testing involved stability, control, and performance which had the goal of decreasing the aircraft's vulnerability to ground fire.

In 1973, there was the final modification to a C-130, in this case a prototype AC-130H with the PAVE SPECTRE II Program. The main purpose of this program was to evaluate the effects of side firing on engine fairings. The flight test program was completed in January 1973 with a total of fifty flight hours.

It was determined that the installation of engine fairings had no noticeable effect on the stability and control of the aircraft. Various store configurations did not affect drag counts with or without fairings.

B-57-based Program
In 1969, there was a side-firing program using a B-57 aircraft. The so-called PAVE GATE Program used a Gatling gun and turret mounted on the B-57. Also mounted was a low light TV sensor.

Testing involved recording ground targets, ground strikes and the pod-mounted fire control system data to demonstrate technical and engineering capabilities during flight evaluation. Acceptance tests were completed in November 1969.

C-131-based Program
During the early 1970s, the 4950th equipped a C-131B outfitted with a television camera housing under the aft fuselage and mounted a laser illuminator in the right wing pod.

The purpose of this installation was to utilize the camera to locate targets on the ground. Initial flight tests showed a strong airframe buffet which resulted from turbulent flow created by the camera housing. The camera had the capability of operating in the absence of any ambient light with the laser illuminating the target area viewed by the camera.

ONE-OF-A-KIND RESEARCH AIRCRAFT

F-15E Strike Eagle Demonstrator Prototype

AIRCRAFT TYPE: YF-15B
MISSION: Numerous test programs along with serving as the F-15E Strike Eagle Demonstrator
TIME PERIOD: Early 1980s through early 1990s
RESPONSIBLE AGENCY: USAF
CONTRACTOR: McDonnell Douglas

THE STORY

It was the second F-15B produced, YF-15B No 71-291, and it certainly saw service in many different research and test capabilities.

With the Y designation, the aircraft was a pre-production version used during the proving of the F-15. Then, it was substantially modified to represent the first F-15E, which although it still carried the F-15 designation, was basically a new aircraft with the name of Strike Eagle. The plane was employed in the fly-off with the F-16XL to determine which one would be produced as a long-range ground attack aircraft. The F-15 won, and a longstanding production of the F-15E was subsequently initiated.

Since all F-15Es were to be dual configurations, it was logical to select either a B or D version of the F-15 for this program, thus the selection of this B version for the testing. The demonstrator aircraft was modified with the advanced fuselage-hugging conformal fuel tanks which would become a standard part of the F-15E. The tanks greatly increased the range of the aircraft, each carrying four hundred gallons of fuel, while producing a minimum of increased drag.

After completing its mission of proving that the F-15E would be an effective ground attack aircraft, the testbed aircraft was used in several other test programs including the "Peak Eagle" program carried out at Wright-Patterson AFB in the early 1990s.

The Strike Eagle Demonstrator took on several looks, shown here in complete camouflage paint. (USAF Photo)

A production F-15B was modified to serve as the demonstrator for the F-15E Strike Eagle shown here with a full air-to-ground ordnance load. (McDonnell Douglas Photo)

PART III: PROTOTYPES

YF-17 as F-18 Prototype

AIRCRAFT TYPE: YF-17
MISSION: To serve as prototype for F-18 Hornet
TIME PERIOD: 1970s
RESPONSIBLE AGENCY: U.S. Navy
CONTRACTOR: Northrop/McDonnell Douglas

THE STORY

It's an interesting story, an aircraft that started life as a prototype for an Air Force program. Losing that competition, it then served as the prototype for a successful Navy program.

Initially, it was the Lightweight Fighter Program in the early 1970s when the Northrop YF-17 was in competition with the General Dynamics YF-16 in a vigorous battle. The competition was tight, but the General Dynamics entry proved to be the winner. The stakes, though, were thought to be much higher in that the overall goal of the program was to produce a common USAF/Navy fighter.

That was to be the plan, but the Navy didn't like the choice of the F-16 as the winner from its point-of-view. So what were the reasons for the Navy's thinking? In a report, the Navy stated that naval air combat missions were much different from those being assigned to the F-16, especially those missions from carriers. The Navy felt that the changes required to the F-16 derivative would be quite extensive, and the capabilities of that design would be greatly compromised. There was also the longtime Navy preference for a multi-engine fighter, which the YF-17 was, and the F-16 wasn't!

The overall goal of the aircraft commonality between the services was, of course, to save money. The Navy was saying that saving money to produce a fighter it didn't want was, in reality, no savings! In the long run, it would convince the powers that it really needed a variant of the YF-17. It, obviously, would become the F-18. In retrospect, it would be a good decision because the F-18 has proved its worth many times over. In fact, in the mid-1990s, there were still advanced versions of the craft (the F/A-18E/F) under consideration.

An artist's concept of the F-18 Hornet fighter that would evolve from the YF-17 testbed prototype. (U.S. Navy Drawing)

The Navy took great advantage of much of the YF-17 flight test data which had been accumulated during the Air Force competition. Then, the YF-17 prototype (with Navy markings, of course), continued to be used in flight testing in the completely new F-18 mode.

It was difficult to ascertain, with its new paint scheme, that this was actually the original YF-17 prototype. The plane was used specifically in the development of the carrier and land based versions of the F-18. In addition, the plane was used in the vigorous effort to sell the F-18 worldwide. For example, the plane flew at the 1976 Farnborough Air Show, the first time the plane had been flown outside the United States.

The performance and mission capabilities for the new fighter would be vigorous, but as is well known, the F-18 would prove to be the Naval weapon of the 1980s and 1990s. It has met or exceeded its initial design goals of Mach 1.5, a radius of action of more than 400 nautical miles, and a combat ceiling in excess of 45,000 feet.

It was an interesting lifetime for the YF-17, which in effect was a prototype for two fighter programs, one it lost and one it won.

The YF-17 served as a prototype during competition with the YF-16 in the Air Force Lightweight Fighter competition. The YF-17 aircraft lost that competition, but it would then serve as the testbed for the Navy F/A-18 fighter. (USAF Photo)

ONE-OF-A-KIND RESEARCH AIRCRAFT

A-37 Prototype

TYPE OF AIRCRAFT: T-37 Tweety Bird Trainer
MISSION: Serve as prototype for development of A-37 attack aircraft
TIME PERIOD: 1960s
RESPONSIBLE AGENCY: USAF Aeronautical Systems Division
CONTRACTOR: Cessna

THE STORY

The trainer aircraft seemed like a strange choice to be modified into a low-altitude attack aircraft, but that is exactly what was done to a T-37 trainer.

The original production version of the T-37 basic trainer carried a pair of Continental J69-T-25 turbojet engines, which each have 1,025 pounds of thrust, providing the aircraft with a 350 miles-per-hour capability and a ceiling of 35,000 feet. The plane carried a crew of two with a maximum take-off weight of 6,580 pounds.

During the course of its long career, the plane would be modified several times in its training configuration. The T-37A made its first flight in 1955, but did not go into service with the Air Force until 1957. All T-37As were eventually modified into the B configuration. The T-37B joined the force in 1959 and incorporated more powerful engines and new radio equipment. The model was flown by a number of Third World countries.

The T-37C followed and incorporated provisions for both armament and wingtip fuel tanks. For that modification, a T-37B production aircraft (No 62-5951) served as the prototype.

But it still continued, obviously into the T-37D, and that is where 62-5951, and another testbed (No 62-5950) were modified into the so-called YAT-37D-CE configuration. The "CE" in that configuration stood for Counter-Insurgency which was definitely the mission of this model. The model incorporated six underwing pylons and a pair of 2400 pound thrust J85-GE-5 powerplants.

The evolution continued as the prototypes were then redesignated YA-37A-CE, and then served as the testbeds for the A-37 Dragonfly aircraft with additional improvements incorporated. The definitive production version which evolved was the A-37B which appeared in September 1967. 577 of the model were built. One of the prototypes was on static display at Lockland AFB, Texas in the early 1970s.

A T-37, such as the one shown here, served as a prototype for several different attack versions of this aircraft. (USAF Photo)

Also from the publisher

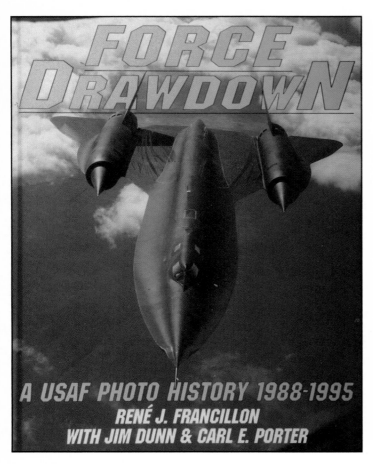
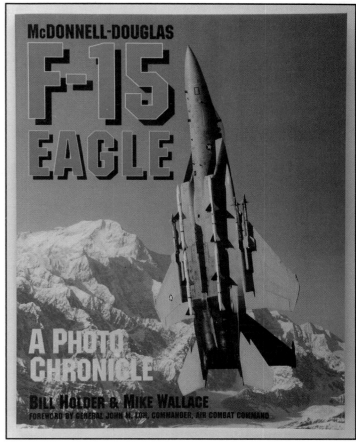

FORCE DRAWDOWN
A USAF Photo History 1988-1995
René J. Francillon
with Jim Dunn & Carl E. Porter

Illustrated with over 410 color photos, this new book provides a rich pictorial record of aircraft (including old and new markings) and units which no longer exist, and offers a visual chronicle of organizational changes between 1988 and 1995.
Size: 8 1/2" x 11" over 410 color photographs
144 pages, hard cover
ISBN: 0-88740-777-3　　　　　　　　　　　　　　　$29.95

McDONNELL-DOUGLAS
F-15 EAGLE
A Photo Chronicle
Bill Holder & Mike Wallace

Photo chronicle covers the F-15 Eagle from planning and development, to its success in Operation Desert Storm and post-Desert Storm. All types are covered, including foreign – Israel, Japan and Saudi Arabia.
Size: 8 1/2" x 11" over 150 color & b/w photographs
88 pages, soft cover
ISBN: 0-88740-662-9　　　　　　　　　　　　　　　$19.95